Surveys of Modern Mathematics
Volume VII

Introductory Lectures on Manifold Topology: Signposts

Thomas Farrell
Department of Mathematical Sciences
Binghamton University

Yang Su
Academy of Mathematics and Systems Science
Chinese Academy of Sciences

International Press
www.intlpress.com

高等教育出版社
HIGHER EDUCATION PRESS

Surveys of Modern Mathematics, Volume VII
Introductory Lectures on Manifold Topology: Signposts

Thomas Farrell
Department of Mathematical Sciences, Binghamton University
Yang Su
Academy of Mathematics and Systems Science, Chinese Academy of Sciences

ISBN: 978-1-57146-287-9

Printed in the United States of America.

18 17 16 15 14   1 2 3 4 5 6 7 8 9

SURVEYS OF MODERN MATHEMATICS

Dedicated to my mother with deepest gratitude

– Y. S.

Preface

The purpose of this book is to introduce to advanced graduate students and other interested mathematicians some of the basic technique and results from manifold topology. It is assumed that the reader is familiar with algebraic topology through cup products and Poincaré duality as well as with fiber bundles and characteristic classes; e.g. with the material in the first half of the book "Characteristic Classes" by J. W. Milnor and J. D. Stasheff. A glance at the Contents shows the topics that are covered. The book is based on a course of lectures given by the first author during the fall semester, 2009 at the Morningside Center of the Chinese Academy of Sciences. It was originally planned as a year long course; hence some of the topics alluded to in the Introduction are not covered here. These will be done in a second volume.

The writing of this book was partially supported by a grant from the National Science Foundation of the USA and by a Visiting Professorship at the Chinese Academy of Sciences of the first author, and by a grant from the National Science Foundation of China of the second author.

Contents

1

Introduction

The book is devoted to the basic problem of classifying manifolds up to homeomorphism and up to diffeomorphism. Today this classification is pretty well understood for many manifolds of geometric significance, including for example those supporting a Riemannian metric of non-positive sectional curvature (except in dimension 4). The starting point for this classification is due to the work of Kervaire, Milnor and Smale from about 1956—1963. Their work can be thought of as starting to answer the following two questions.

Question 1.1 (*Kervaire-Milnor*). Given a topological manifold M^n (of dimension n), does M^n support a differential structure? And if so, how many non-diffeomorphic structures does M^n support?

Question 1.2 (*Smale*). Given a topological space K, does there exist a compact (topological) manifold (without boundary) homotopy equivalent to K? And if so, how many non-homeomorphic such manifolds are there?

We begin with some explanatory definitions.

An n-dimensional (topological) manifold is a Hausdorff topological space M which is locally homeomorphic to n-dimensional Euclidean space $\mathbb{R}^n$. Additionally one requires the topology on M to have a countable basis; i.e. there exists a countable collection $\{U_n\}$ of open subsets U_n of M such that any open set U in M is a union of sets from this collection. (When M is compact, this condition is superfluous. And we will be mainly interested in compact manifolds which are also called *closed manifolds.*)

But the original reason for thinking about manifolds (probably due to Riemann) was as objects to do calculus on. For this the notion of smooth manifold is needed. And perhaps the most conceptual way to describe a (compact) smooth m-dimensional manifold is as a closed and bounded subset of some Euclidean space $\mathbb{R}^n$ ($n \gg m$) which has an m-dimensional tangent plane T_x at each point $x \in M$; i.e. there exists a pair of perpendicular affine subspaces T_x, N_x passing through x such that

$$\mathbb{R}^n = T_x \times N_x$$

and M is locally the graph of a smooth function

$$\varphi_x \colon U_x \to N_x$$

where $(D\varphi_x)_{x=0} = 0$. (Here U_x is an open neighborhood of $x \in T_x$.)

A smooth m-manifold is clearly a topological m-manifold; but it has more structure. Namely a cover by open sets $V_x = \operatorname{graph}(\varphi_x)$ and *charts* $\psi_x \colon V_x \to T_x = \mathbb{R}^m$ where ψ_x is the perpendicular projection into T_x. These charts have the following two properties:

1. $\psi_x \colon V_x \to U_x \subseteq T_x = \mathbb{R}^m$ is a homeomorphism;
2. $\psi_x \circ \psi_y^{-1} \colon \psi_y(V_x \cap V_y) \to T_x = \mathbb{R}^m$ is a smooth map.

That is to say, the charts $\{(V_x, \psi_x)|\ x \in M\}$ are a *smooth structure* on M. This is the definition of a smooth manifold given in textbooks; i.e. a smooth manifold is a topological manifold equipped with a smooth structure. Whitney's famous embedding theorem showed that every smooth structure on a topological manifold M arises in the manner described above.

A continuous map $f \colon M \to N$ between a topological m-manifold M and a topological n-manifold N equipped with smooth structures Φ_M and Φ_N, respectively, is smooth if for each pair of charts $(V, \varphi) \in \Phi_M$ and $(U, \psi) \in \Phi_N$, the composite map

$$\psi \circ f \circ \varphi^{-1} \colon \mathbb{R}^m \supseteq \varphi(V \cap f^{-1}(U)) \to \psi(U) \subseteq \mathbb{R}^n$$

is smooth (in the calculus sense).

In particular smooth structures Φ and Ψ on the *same* topological manifold M are said to be *equivalent* provided both $\operatorname{id} \colon (M, \Phi) \to (M, \Psi)$ and $\operatorname{id} \colon (M, \Psi) \to (M, \Phi)$ are smooth maps.

Example 1.3. Consider $M = \mathbb{R}$, with smooth structures

$$\Phi = \{(\mathbb{R}, \varphi(x) = x)\}, \quad \Psi = \{(\mathbb{R}, \psi(x) = x^3)\}.$$

Now $\operatorname{id} \colon (M, \Phi) \to (M, \Psi)$ is smooth; but $\operatorname{id} \colon (M, \Psi) \to (M, \Phi)$ is *not*. Therefore Φ and Ψ are inequivalent smooth structures on $\mathbb{R}$.

Hence even the simplest topological manifold supports inequivalent smooth structures. But there is a weaker equivalence relation between smooth structures which is one of the primary interest in differential topology; namely that of diffeomorphic structures.

A homeomorphism $f \colon (M, \Phi) \to (N, \Psi)$ between topological manifolds M and N equipped with smooth structures Φ and Ψ, respectively, is a *diffeomorphism* provided both

$$f \colon (M, \Phi) \to (M, \Psi)$$

and

$$f^{-1} \colon (M, \Psi) \to (M, \Phi)$$

are smooth. When a diffeomorphism exists we say that the two differentiable manifolds (M, Φ) and (N, Ψ) are diffeomorphic. This is the weaker equivalence relation mentioned above.

Note that the two differential structures Φ and Ψ on the manifold $\mathbb{R}$ in the above example are diffeomorphic even though they are inequivalent; in fact $f(x) = x^{1/3}$ is a diffeomorphism $f\colon (\mathbb{R}, \Phi) \to (\mathbb{R}, \Psi)$.

Sixty years ago it was generally assumed that every topological manifold M supports a differential structure and that the structure is *unique* up to diffeomorphism. Furthermore this belief had been recently strengthened by the proof of Hilbert's 5th Problem (Gleason, Montgomery-Zippen) which implied in the case that M is a topological group, then M supports a smooth structure such that both multiplication: $M \times M \to M$ and inverse: $M \to M$ are smooth maps. Furthermore any two smooth structures on M satisfying these conditions are diffeomorphic.

But in 1956 Milnor startled the mathematical community by constructing an "exotic" differential structure Ψ on the 7-dimensional sphere $S^7 = \{x \in \mathbb{R}^8 | \, |x| = 1\}$. S^7 has a canonical differential structure Φ determined by its "birth certificate" embedding in $\mathbb{R}^8$, as explained above. But Milnor's structure comes from some other subspace $\Sigma \subseteq \mathbb{R}^{15}$ having 7-dimensional tangent planes T_x, $x \in \Sigma$, again as explained above. There is a homeomorphism $f\colon S^7 \to \Sigma$; but *no* diffeomorphism. Formally Ψ is the pull to S^7 via f of the structure on Σ induced by its embedding $\Sigma \subseteq \mathbb{R}^{15}$. And four years later Kervaire constructed a 10-dimensional topological manifold which does *not* support any differential structure.

To state the following conjectures we need more explanatory definitions.

A continuous map $f\colon X \to Y$ between topological spaces X and Y is a *homotopy equivalence* provided there exists a second continuous map $g\colon Y \to X$ such that both $g \circ f$ is homotopic to id_X (written $g \circ f \sim \mathrm{id}_X$) and $f \circ g \sim \mathrm{id}_Y$. And such a map g (if it exists) is called a homotopy inverse to f. A homotopy inverse is unique up to homotopy.

We say that two topological spaces X and Y are homotopy equivalent if there exists a homotopy equivalence $f\colon X \to Y$. Since every homeomorphism is clearly a homotopy equivalence, homotopy equivalence puts a weaker equivalence relation on the collection of all topological spaces than homeomorphism; i.e. if X and Y are homeomorphic then they are homotopy equivalent. But algebraic topology tools are quite effective for determining when two spaces are homotopy equivalent. Hence the following conjecture:

Naive Conjecture 1.4. *Homotopy equivalent closed* (*i.e. compact*) *manifolds are homeomorphic.*

This conjecture is true for 2-dimensional manifolds. However in the 1930s some simple natural examples were exhibited showing this Naive Conjecture to be false even for 3-dimensional manifolds. But only non-simply connected examples were known. This prompted Hurewicz to make the following conjecture.

Hurewicz Conjecture 1.5. *Simply-connected homotopy equivalent closed manifolds are homeomorphic.*

Remark 1.6. This well-known conjecture is explicitly stated and discussed as Problem 31 on the list compiled by R. Lashof , "Problems in differential and algebraic topology", Seattle conference, 1963, Ann. of Math. 81 (1965), 565–591.

His conjecture was a vast generalization of Poincaré's conjecture that every simply-connected closed 3-dimensional manifold M^3 is homeomorphic to the 3-sphere S^3. To see this one of course needs only show that M is homotopy equivalent to S^3. But this follows from Poincaré duality, Hurewicz Theorem and Whitehead Theorem. We briefly sketch the argument since similar arguments will frequently occur. First of all, M^3 is orientable, because its orientation (2-sheeted) covering space is trivial since $\pi_1(M^3) = 1$. By Poincaré duality

$$H_2(M^3) \cong H^1(M^3) = \mathrm{Hom}(\pi_1 M, \mathbb{Z}) = 0.$$

By Hurewicz Theorem $\pi_3(M^3) = H_3(M^3) = \mathbb{Z}$. Let $f\colon S^3 \to M^3$ generate $\pi_3(M^3)$. Then f is a homotopy equivalence by Whitehead Theorem which asserts that any map between simply-connected "nice spaces" which induces an isomorphism between their homology groups is a homotopy equivalence. And manifolds are particular instances of nice spaces.

2

The h-Cobordism Theorem

2.1 The h-Cobordism Theorem and Generalized Poincaré Conjecture

Smale proved a very important special case of the Hurewicz Conjecture in 1961. Namely he showed that any closed (smooth) manifold M homotopy equivalent to the m-sphere S^m, where $m \geq 6$, is homeomorphic to S^m. This verified what had been known as the Generalized Poincaré Conjecture (GPC).

Remark 2.1. Stallings (1965) improved Smale's result to include $m = 5$, and Newman (1966) showed the assumption that M has a smooth structure is superfluous. They both used a technique much different from Smale's, called engulfing.

The fundamental result from which the Generalized Poincaré Conjecture follows is Smale's h-Cobordism Theorem, which we proceed to formulate. For this we need to generalize the notion of m-dimensional manifold to m-dimensional manifold with boundary.

A topological m-dimensional manifold with boundary is again a Hausdorff topological space W whose topology has a countable basis, but which is now assumed to be locally homeomorphic to either $\mathbb{R}^m$ or $[0,\infty) \times \mathbb{R}^{m-1}$; i.e. each point $x \in W$ has an open neighborhood homeomorphic to either $\mathbb{R}^m$ or $[0,\infty) \times \mathbb{R}^{m-1}$. Those points $x \in W$ which have no neighborhood homeomorphic to $\mathbb{R}^m$ form the boundary ∂W of W. It can be shown that either $\partial W = \emptyset$ or ∂W is an $(m-1)$-dimensional manifold (without boundary; i.e. $\partial(\partial W) = \emptyset$) and $W - \partial W$ is an m-dimensional manifold. Furthermore the collar neighborhood theorem says that ∂W has an open neighborhood C, "a collar of ∂W in W" such that C is homeomorphic to $\partial W \times [0,\infty)$ and via a homeomorphism

$$f\colon \partial W \times [0,\infty) \to C \subseteq W$$

such that $f(x,0) = x$.

A differential structure Φ on a topological m-manifold W with boundary is again defined by giving a collection of charts

$$\psi_x\colon V_x \to U_x$$

for each $x \in W$, as before in defining a smooth structure on a topological m-manifold (without boundary), except now U_x is an open subset of $[0, \infty) \times \mathbb{R}^{m-1}$.

If Φ is a smooth structure on a topological m-manifold W with boundary then ∂W inherits, in a natural way, a smooth structure $\Phi_{\partial W}$ from Φ. And the collar map

$$f \colon \partial W \times [0, \infty) \to C$$

can be chosen to be a diffeomorphism.

Example 2.2. $\mathbb{D}^n = \{x \in \mathbb{R}^n | \ |x| \leq 1\}$ is a smooth n-dimensional manifold with $\partial \mathbb{D}^n = S^{n-1}$. A collar C in this case can be taken to be $\mathbb{D}^n - 0$. We refer to $\mathbb{D}^n$ as the "n-disc" or "n-ball".

A *cobordism* is a manifold W with boundary where ∂W is expressed as the disjoint union of two non-empty open subsets $\partial^- W$ and $\partial^+ W$ (which are now both $(\dim W - 1)$-dimensional manifolds); W is said to be a cobordism from $\partial^- W$ to $\partial^+ W$.

Example 2.3. Here is a 2-dimensional example.

In the pictured example (Figure 2.1), two small (open) discs have been deleted from the 2-dimensional torus $T^2 = S^1 \times S^1$.

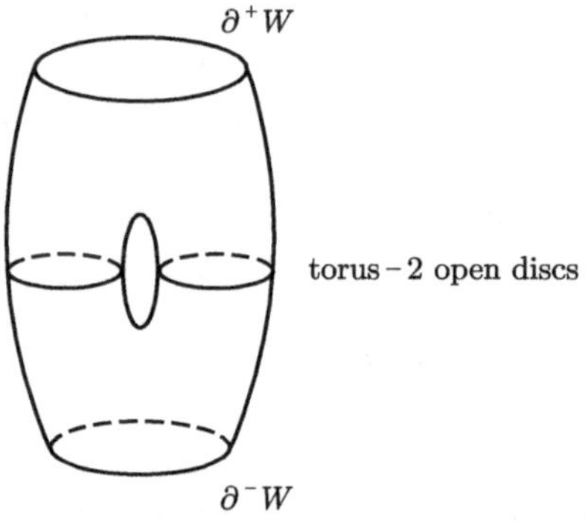

Figure 2.1

Definition 2.4. *An* h-cobordism *is a compact cobordism* W *such that both inclusion maps* $\partial^- W \subseteq W$ *and* $\partial^+ W \subseteq W$ *are homotopy equivalences.*

The cobordism W in the example above is not an h-cobordism since $\pi_1(\partial^- W)$ and $\pi_1(W)$ are not isomorphic; $\pi_1(\partial^- W) \cong \mathbb{Z}$ is an abelian group while $\pi_1(W) \cong \mathbb{Z} * \mathbb{Z} * \mathbb{Z}$ is not abelian.

But there is a canonical example of an h-cobordism; namely $W = M \times [0, 1]$, the cylinder with base a closed manifold M (Figure 2.2).

It is hard to think of another example. In fact that is the gist of Smale's Theorem.

***h*-Cobordism Theorem.** *Let* W^m *be a smooth* h*-cobordism. If* $\pi_1(W) = 1$ *and* $m \geq 6$*, then* W *is diffeomorphic to* $\partial^- W \times [0, 1]$.

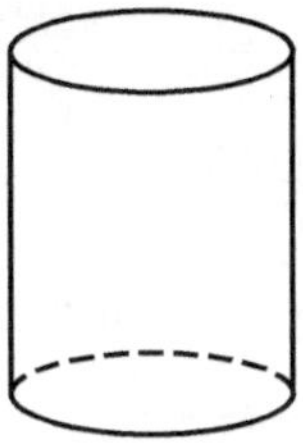

Figure 2.2

Addendum 2.5. *The diffeomorphism* $f\colon \partial^- W \times [0,1] \to W$ *can be chosen so that* $f(x,0) = x$.

We now give Smale's deduction of the Generalized Poincaré Conjecture (GPC) from his h-Cobordism Theorem. Let D^+ and D^- be a pair of disjoint m-dimensional balls nicely embedded in M (say inside domains of disjoint charts) as shown in Figure 2.3. Let W be M with $\mathring{D}^+ \cup \mathring{D}^-$ deleted. Here $\mathring{D}^+$ and $\mathring{D}^-$ are the interiors of D^+ and D^- respectively. It is easy to see that W is a smooth cobordism from $\partial^- W$ to $\partial^+ W$ where $\partial^- W = \partial D^-$ and $\partial^+ W = \partial D^+$.

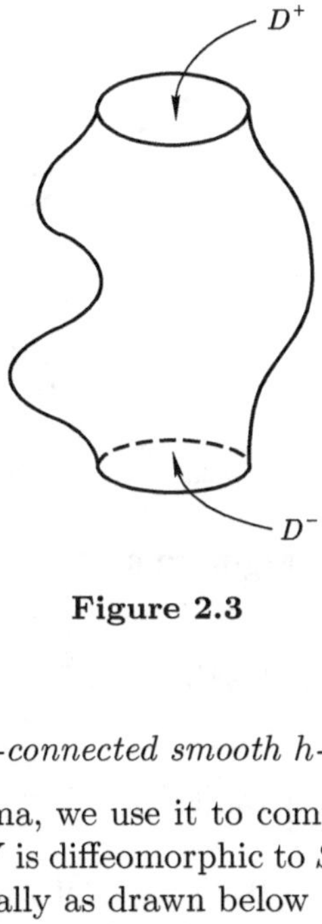

Figure 2.3

Lemma 2.6. *W is a simply-connected smooth h-cobordism.*

Before verifying this lemma, we use it to complete Smale's proof of GPC. By his h-Cobordism Theorem, W is diffeomorphic to $S^{m-1} \times [0,1]$ since $\partial D^- = S^{m-1}$; hence the picture above is really as drawn below in Figure 2.4.

The only ambiguity is how the three known pieces in this picture fit together. By Addendum 2.5, the bottom piece fits canonically into the middle piece; hence $W \cup D^-$ can be identified with $\mathbb{D}^m$. Therefore M^m is diffeomorphic to $\mathbb{D}^m \cup_f \mathbb{D}^m$ where $f\colon S^{m-1} \to S^{m-1}$ is a self diffeomorphism; i.e. M^m is a twisted double of $\mathbb{D}^m$ (Figure 2.5).

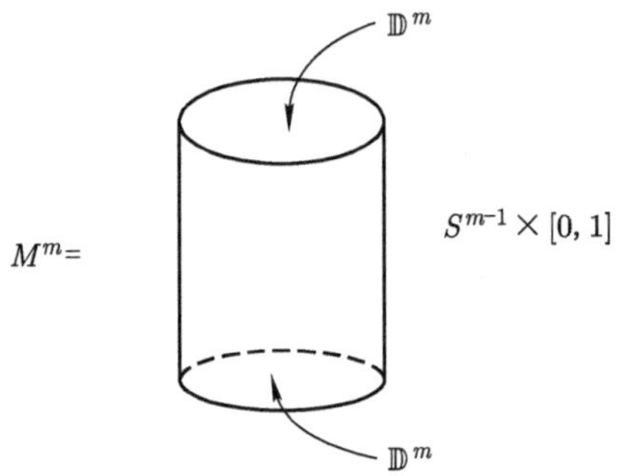

Figure 2.4

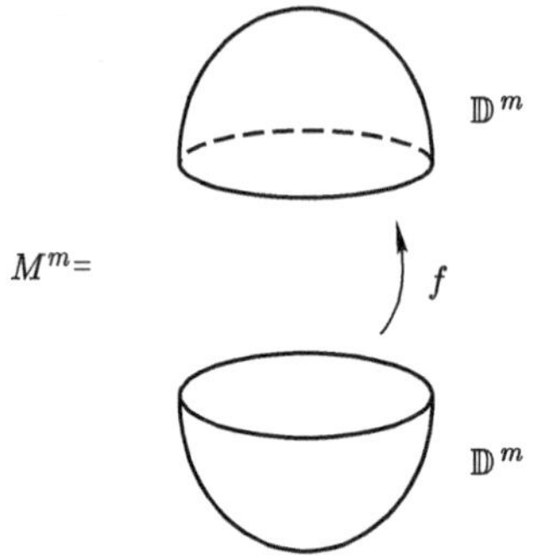

Figure 2.5

Remark 2.7. This is a quite important fact of independent interest which is fundamental to the Kervaire-Milnor analysis of "exotic spheres" to be described later.

Note that if $f = \mathrm{id}$, then $\mathbb{D}^m \cup_{\mathrm{id}} \mathbb{D}^m = S^m$. But we know, by the result of Milnor in 1956 mentioned in Chapter 1, that M^m is in general *not* diffeomorphic to S^m. But we can construct a homeomorphism $S^m \to M^m$ as illustrated below in Figure 2.6.

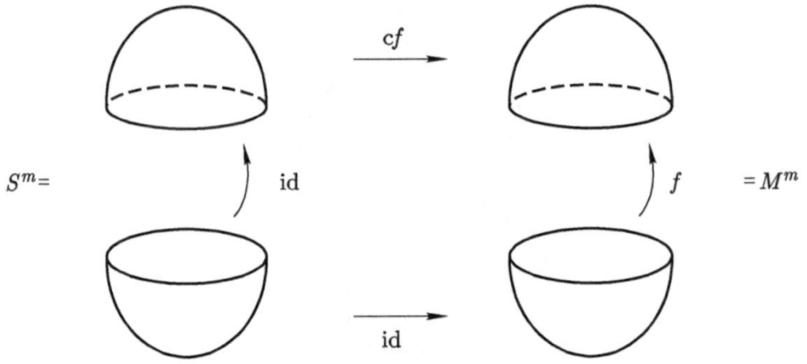

Figure 2.6

In this illustration $cf\colon \mathbb{D}^m \to \mathbb{D}^m$ is the "cone" of f which is defined analytically by

$$cf(x) = \begin{cases} |x| f(\frac{x}{|x|}), & \text{if } x \neq 0, \\ 0, & \text{if } x = 0 \end{cases}$$

where $x \in \mathbb{D}^m$; i.e. $|x| \leq 1$. But more informative is its pictorial description below in Figure 2.7.

Thus the proof of GPC is complete once Lemma 2.6 is verified.

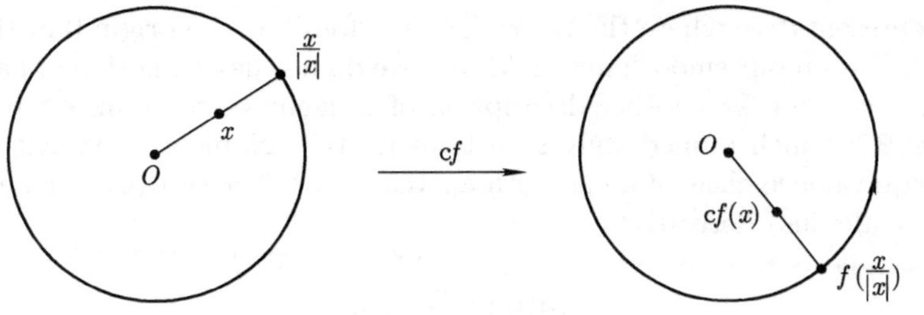

Figure 2.7

Proof of Lemma 2.6. Let $\mathscr{D}^+ = D^+ \cup W$ and $\mathscr{D}^- = D^- \cup W$. They are both m-manifolds with boundary. We proceed to calculate the reduced homology groups $\widetilde{H}_i(\mathscr{D}^+)$. For this purpose we apply the reduced Mayer-Vietoris sequence to the decomposition $M^m = \mathscr{D}^+ \cup D^-$, $\mathscr{D}^+ \cap D^- = S^{m-1}$. After observing that $\widetilde{H}_*(D^-) = 0$, we obtain

$$\cdots \to \widetilde{H}_i(S^{m-1}) \to \widetilde{H}_i(\mathscr{D}^+) \to \widetilde{H}_i(M^m) \to \cdots .$$

From this exact sequence we immediately deduce that

$$\widetilde{H}_i(\mathscr{D}^+) = 0 \quad \text{for } i \neq m, m-1.$$

To calculate the remaining two groups we observe, using Lefschetz duality, that

$$H_m(\mathscr{D}^+) = H^0(\mathscr{D}^+, S^{m-1}), \quad H_{m-1}(\mathscr{D}^+) = H^1(\mathscr{D}^+, S^{m-1}).$$

And these groups are both zero as is seen by applying the long exact sequence in cohomology to the pair $(\mathscr{D}^+, S^{m-1})$. Consequently $\widetilde{H}_i(\mathscr{D}^+) = 0$ for all i. A completely analogous argument yields that $\widetilde{H}_i(\mathscr{D}^-) = 0$ for all i. And a similar but easier argument using Van Kampen Theorem in place of the Mayer-Vietoris sequence yields that both $\mathscr{D}^+$ and $\mathscr{D}^-$ are simply connected. Another application of Van Kampen Theorem to the decomposition $\mathscr{D}^+ = W \cup D^+$, $W \cap D^+ = \partial^+ W = S^{m-1}$ yields that W is simply connected. Next applying the Mayer-Vietoris sequence to the decomposition $\mathscr{D}^+ = W \cup D^+$, $W \cap D^+ = \partial^+ W$ yields that the inclusion map $\partial^+ W \subseteq W$ induces an isomorphism on homology groups, and hence by Whitehead Theorem (mentioned in Chapter 1) it is a homotopy equivalence. A completely analogous argument shows that the other inclusion $\partial^- W \subseteq W$ is also a homotopy equivalence. This completes the verification of Lemma 2.6, and hence the relation of GPC to the h-Cobordism Theorem. □

The first major project of this course is to sketch a proof of Smale's h-Cobordism Theorem. (A detailed proof is the subject of Milnor's book [13].) We start our sketch with some more basic definitions in differential topology.

2.2 Tangent vectors, embeddings, isotopies

For $\mathbb{R}^m$ we have the standard calculus notion of a tangent vector; i.e. an equivalence class of parallel arrows all having the same length. And each tangent vector has a unique representative where the arrow has its "foot" at the origin 0 of the vector space $\mathbb{R}^m$. But on our smooth manifold M^m we do not have a notion of parallelism like in $\mathbb{R}^m$; so we take another description of tangent vector v in $\mathbb{R}^m$ based at a point $x \in \mathbb{R}^m$ which immediately translates to M^m. Namely a tangent vector at x is the equivalence class of all smooth curves α in $\mathbb{R}^m$ with $\alpha(0) = x$ and having the same (calculus) derivative

$$\dot{\alpha}(0) := \frac{d\alpha}{dt}|_{t=0}$$

at "time" $t = 0$. Here $\alpha \colon I \to \mathbb{R}^m$, where I is any open interval in $\mathbb{R}$ containing 0. In terms of the above standard definition, this tangent vector is $\dot{\alpha}(0) \in \mathbb{R}^m$. Then we define a tangent vector at $x \in M^m$ to again be an equivalence class of smooth curves α in M^m with $\alpha(0) = x$. Where two such curves α, γ are equivalent if for a co-ordinate chart $(U, \varphi) \in \Phi_M$ with $x \in U$, we have that $\varphi \circ \alpha$ and $\varphi \circ \gamma$ represent the same tangent vector at $\varphi(x) \in \mathbb{R}^m$. We denote the tangent vector that the curve α represents by $\dot{\alpha}(0)$. We denote the set of all vectors tangent to M^m at x by $T_x M^m$. It has a natural vector space structure; e.g. $\lambda\dot{\alpha}(0) = \dot{\gamma}(0)$ where $\gamma(t) = \alpha(\lambda t)$, $t \in \mathbb{R}$.

Remark 2.8. If $M^m \subseteq \mathbb{R}^n$ is an n-manifold as in Chapter 1, then $T_x M^m$ can be canonically identified with the affine subspace T_x mentioned there.

Let $f \colon M^m \to N^n$ be a smooth map, then its derivative at $x \in M$ is the linear transformation

$$df_x \colon T_x M^m \to T_{f(x)} N^n$$

define by

$$df_x(\dot{\alpha}(0)) = \frac{d}{dt} f \circ \alpha(t)|_{t=0}$$

(Figure 2.8).

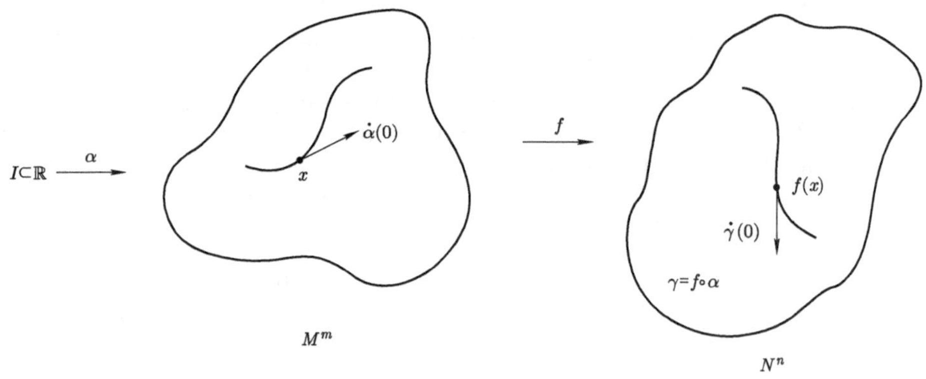

Figure 2.8

(A quite complete reference book on the basics of differential topology is Hirsch's book [7]. Also see the first part of the Milnor-Stasheff book [17].)

An *immersion* $f\colon M^m \to N^n$ is a smooth map such that $df_x\colon T_xM^m \to T_{f(x)}N^n$ is monic for each $x \in M$.

Example 2.9. The map $f\colon \mathbb{R} \to \mathbb{R}$ defined by $f(t) = t^3$ is *not* an immersion since $df_0 = 0$.

An *embedding* $f\colon M^m \to N^n$ is an immersion such that $f\colon M \to f(M) \subseteq N$ is a homeomorphism into its image.

Exercise 2.10. Parametrize the "curve" $\bigvee$ in $\mathbb{R}^2$ by $\alpha\colon \mathbb{R} \to \mathbb{R}^2$ so that α is smooth and is a homeomorphism onto its image; but α is *not* an immersion and hence *not* an embedding.

An *isotopy* is a "smoothly varying" 1-parameter family of embeddings

$$f_t\colon M^m \to N^n, \quad t \in [0,1].$$

"Smoothly varying" means that the function $F\colon M \times [0,1] \to N$ defined by $F(x,t) = f_t(x)$ is smooth. (Here $x \in M$, $t \in [0,1]$.)

Caveat 2.11. *If we do not insist on "smoothly varying" in the definition of isotopy, then knots in $\mathbb{R}^3$ can be* unknotted *by an isotopy. Just pull them tight. That this cannot happen with the smoothly varying condition is a consequence of the following:*

Isotopy Extension Theorem. *If $f_t\colon M \to N$ is an isotopy, then there exists a self-isotopy $F_t\colon N \to N$ such that*

1. $F_0 = \mathrm{id}_N$*;*
2. $f_t = F_t \circ f_0$ *for all* $t \in [0,1]$.

This important result is a consequence of the fundamental existence and uniqueness theorem for ordinary differential equation (ODE). Its proof is quite conceptual and is found in Hirsch's book [7]. We will next sketch it.

Remark 2.12. Because of Condition 2, the diffeomorphism $F_1\colon N \to N$ maps $f_0(M)$ to $f_1(M)$ and hence $N - f_0(M)$ is diffeomorphic to $N - f_1(M)$. Hence a knot in $\mathbb{R}^3$ cannot be unknotted by an isotopy. We recall that a knot is unknotted if and only if its complement is simply connected.

A vector field X on M^m is an assignment to each point $x \in M$ of a tangent vector $X_x \in T_xM$. If M^m is $\mathbb{R}^m$ we can identify a vector field with a function $X\colon \mathbb{R}^m \to \mathbb{R}^m$; X is said to be continuous (or smooth) provided this function is continuous (or smooth). In the general case a vector field X on M^m is continuous (or smooth) if all its co-ordinate chart representations are continuous (or smooth). Recall that we have defined $\dot\alpha(0) \in T_{\alpha(0)}M^m$ to be the tangent vector represented by a smooth curve $\alpha\colon I \to M$. We extend this to define $\dot\alpha(t_0) \in T_{\alpha(t_0)}M^m$, $t_0 \in I$, to mean $\dot\gamma(0)$ where $\gamma(t) = \alpha(t + t_0)$. By ODE on M determined by the vector field X, we mean the equation

$$\dot\alpha(t) = X_{\alpha(t)}. \tag{2.1}$$

A solution α to (2.1) is a curve that is everywhere "tangent" to the vector field X, and a solution to (2.1) is called an *integral curve* to X.

Fundamental Theorem of ODE. *Let X be a smooth vector field on a closed manifold M^m and $x \in M$. Then there exists a unique integral curve α_x to X satisfying the initial condition $\alpha_x(0) = x$. Furthermore α_x varies smoothly with x; i.e. the function $F\colon M \times \mathbb{R} \to M$ defined by $F(x,t) = \alpha_x(t)$ is smooth. And F defines a flow on M. That is, define a function $f_t\colon M \to M$, for each $t \in \mathbb{R}$, by the equation*

$$f_t(x) = F(x,t) = \alpha_x(t).$$

Then $f_{s+t} = f_s \circ f_t$; hence each f_t is a diffeomorphism. And $t \mapsto f_t$ is a (continuous) group homomorphism from the additive group of $\mathbb{R}$ to the group $\mathrm{Diff}(M)$ *of all diffeomorphisms of M. And the orbits of this flow are the integral curves to X.*

Remark 2.13. This result follows from the basic calculus existence and uniqueness theorem for a system of n first order ODE's by analyzing one co-ordinate chart at a time. (See Milnor's book [14, pp. 9–11] for more details.)

Proof of Isotopy Extension Theorem. (key ideas)

The strategy is to use a particular vector field X on $N \times [0,1]$ from among all of those satisfying the following side condition: For each $(x,t) \in N \times [0,1]$,

$$X_{(x,t)} = Y_{(x,t)} + \frac{d}{dt} \tag{2.2}$$

where $Y_{(x,t)} \in T_x N$ and $\frac{d}{dt}$ is the canonical unit length tangent vector to $\mathbb{R}$ which points "upwards". Note that $T_{(x,t)}(N \times [0,1]) = T_x X \oplus T_t \mathbb{R}$. Although $N \times [0,1]$ is not closed since it has a boundary, the integral curves $\alpha_{(x,0)}$ to X still have the following properties:

1. Their domain is $[0,1]$ and the second component of $\alpha_{(x,0)}$ is t, for each $t \in [0,1]$.
2. Let $F_t(x)$ denote the first component of $\alpha_{(x,0)}$, then $F_t\colon N \to N$ is a diffeomorphism. And $t \mapsto F_t$ is a self-isotopy of N such that $F_0 = \mathrm{id}_N$.

So it remains to choose among these vector fields one such that $f_t = F_t \circ f_0$ for all $t \in [0,1]$. For this consider the codimension-1 submanifold $\mathcal{M}$ of $N \times [0,1]$ defined by

$$\mathcal{M} = \{(f_t(x), t) | (x,t) \in M \times [0,1]\}.$$

The tangent vectors to the family of curves $\{\gamma_x | x \in M\}$, where $\gamma_x(t) = (f_t(x), t)$, define a vector field X on $\mathcal{M}$ with integral curves

$$\alpha_{(f_0(x),0)}(t) = \gamma_x(t) = (f_t(x), t) \tag{2.3}$$

for all $x \in M$ and $t \in [0,1]$. And a routine argument extends X to all of $N \times [0,1]$ so that Eq. (2.2) is satisfied. Since the first component of $\alpha_{(f_0(x),0)}(t)$ is $f_t(x)$ because of Eq. (2.3), we have by Property 2 above that

$$F_t(f_0(x), t) = f_t(x)$$

for all $x \in M$ and $t \in [0,1]$. □

2.3 Handles and handlebody decomposition

The n-dimensional handle h of index $I(h) = i$ is $\mathbb{D}^i \times \mathbb{D}^{n-i}$. It is a topological manifold with boundary homeomorphic to $\mathbb{D}^n$. And its boundary ∂h is naturally the union of two smooth codimension-0 submanifolds

$$\partial^- h = \partial\mathbb{D}^i \times \mathbb{D}^{n-i} = S^{i-1} \times \mathbb{D}^{n-i}$$

and

$$\partial^+ h = \mathbb{D}^i \times \partial\mathbb{D}^{n-i} = \mathbb{D}^i \times S^{n-(i+1)}$$

which intersect in their common boundary; i.e.,

$$\partial^- h \cap \partial^+ h = \partial(\partial^- h) = \partial(\partial^+ h) = \partial\mathbb{D}^i \times \partial\mathbb{D}^{n-i} = S^{i-1} \times S^{n-(i+1)}.$$

Example 2.14. For the case $n = 3$, $i = 2$ see Figure 2.9.

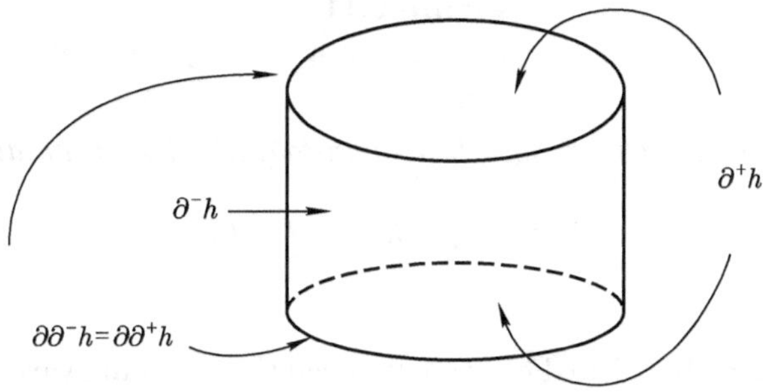

Figure 2.9

Given a smooth n-manifold M^n and an embedding $f\colon \partial^- h \to \partial M$, we can form a new "smooth" n-manifold $M + h$ from the disjoint union of $M \sqcup h$ by glueing $\partial^- h$ to ∂M via f (and then rounding the corners); i.e. identify $x \in \partial^- h$ to $f(x) \in \partial M$ (Figure 2.10).

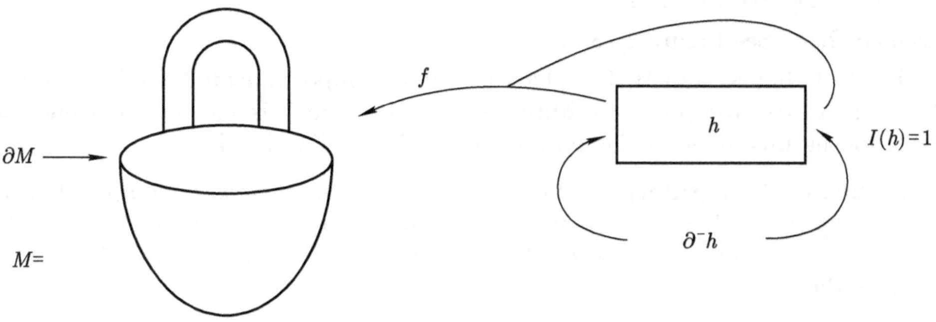

Figure 2.10

The "corner" in Figure 2.10 is "rounded" by attaching only the codimension-0 topological submanifold $\hat{h} \subset h$ to M instead of all of h as illustrated in Figure 2.11 where $\hat{h}$ is the shaded region. Note that $\partial^- h \subseteq \partial \hat{h}$.

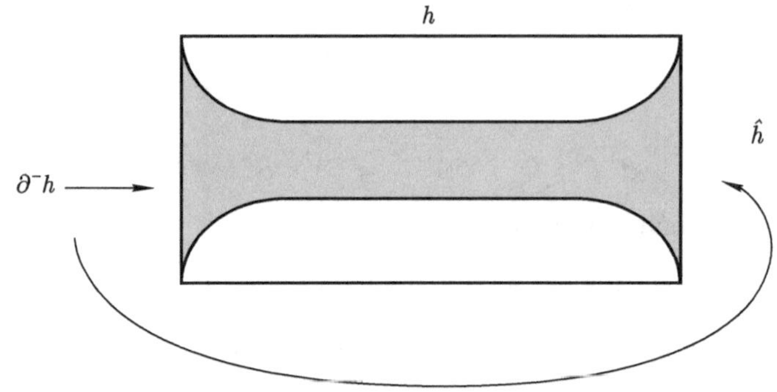

Figure 2.11

Theorem 2.15. *Every closed smooth n-manifold M^n has a handlebody decomposition*

$$M^n = h_1 + h_2 + \cdots + h_m$$

where $I(h_i) \leq I(h_{i+1})$.

To prove the h-Cobordism Theorem, we need the following variant of this result.

Theorem 2.16. *Let W be a smooth n-dimensional cobordism between $\partial^- W$ and $\partial^+ W$. Then*

$$W = \partial^- W \times [0,1] + h_1 + h_2 + \cdots + h_m$$

where $\partial^- W \times [0,1]$ is a collar neighborhood of $\partial^- W$ in W and $\partial^- W \times 0$ is untouched by the handle attaching maps.

Example 2.17. See Figure 2.12.

Example 2.18. See Figure 2.13.

The fact that M and W have handlebody decompositions follows from Morse theory. It is explicitly proved in Milnor's book [14]. But it is also a consequence of the following fundamental theorem due to J. H. C. Whitehead.

Whitehead's Triangulation Theorem. *Every compact smooth manifold M can be smoothly triangulated; i.e. there exists a homeomorphism $f\colon K \to M$ where K is a finite simplicial complex such that for each simplex Δ in K $f|\Delta\colon \Delta \to M$ is an embedding.*

Then a handlebody decomposition for M can be constructed as illustrated in Figure 2.14 (where $M = S^1 \times S^1$).

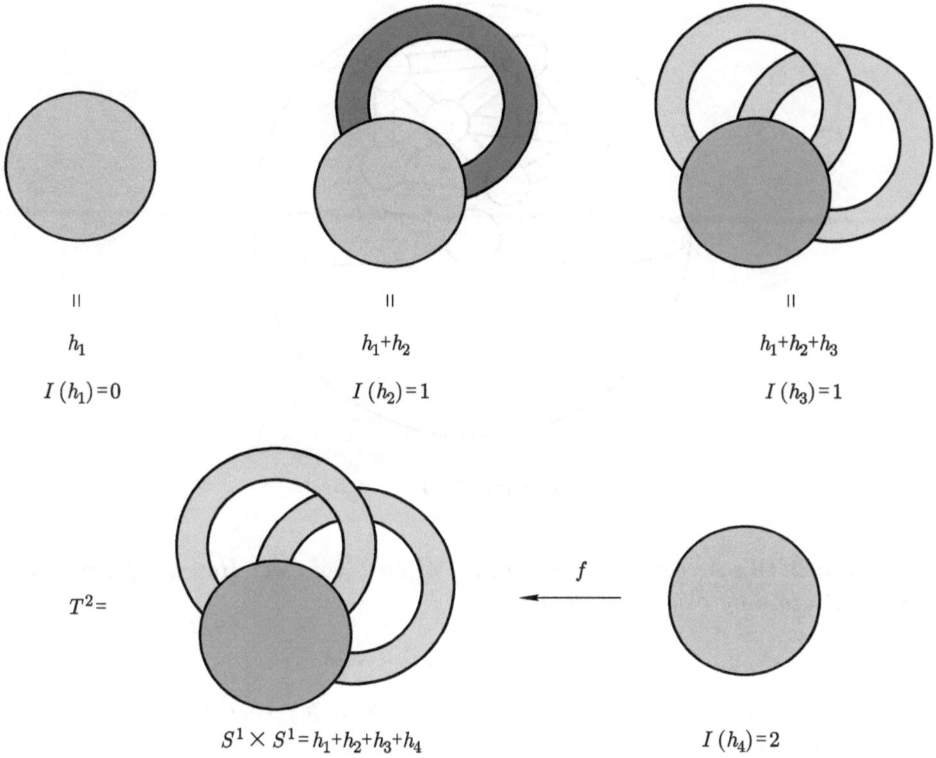

Figure 2.12

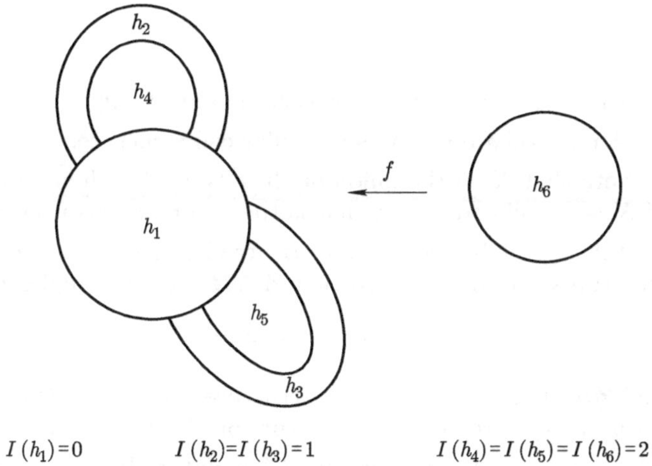

Figure 2.13

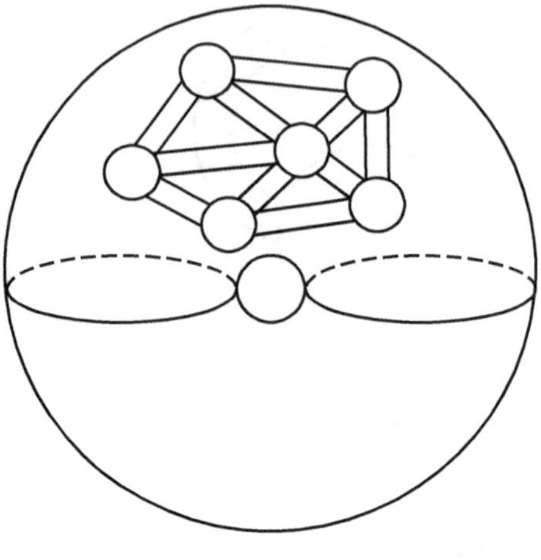

Figure 2.14

Definition 2.19. *A* regular cell structure $\mathscr{C}$ *on a compact Hausdorff space* X *is a finite filtration by closed subspaces*

$$X^0 \subset X^1 \subset \cdots \subset X^n = X$$

(called the skeleta of X*; i.e.* X^i *is the* i*-skeleton of* X*) such that* X^0 *is a finite set; each element of* X^0 *is called a* 0*-cell (or vertex) of the cell structure. And for each* $0 < i \leq n$, $X^i - X^{i-1}$ *is the disjoint union of open subsets of* X^i, σ_1, $\sigma_2, \cdots, \sigma_{m_i}$, *called the (open)* i*-cell of* $\mathscr{C}$*, such that for each of these cells* σ*, there is a homeomorphism*

$$f \colon \mathbb{D}^i \to \overline{\sigma}$$

with $f(\mathrm{Int}\mathbb{D}^i) = \sigma$.

Remark 2.20. A reference on this matter is N. Steenrod [19].

Example 2.21. A triangulation of X is a regular cell structure.

Remark 2.22. Note that X^i is the union of all cells in $\mathscr{C}$ of dimension $\leq i$. And a *subcomplex* of X is any closed subset that is the union of cells from $\mathscr{C}$.

Remark 2.23. If $X = A \cup B$ where A and B are closed subspaces equipped with regular cell structures $\mathscr{C}_A$ and $\mathscr{C}_B$ such that $A \cap B$ is a subcomplex of *both* with

$$A^i \cap B = A \cap B^i = A^i \cap B^i;$$

that is, $\mathscr{C}_A$ and $\mathscr{C}_B$ induce the same cell structure on $A \cap B$. Then $\mathscr{C}_A$ and $\mathscr{C}_B$ induce, in a natural way, a regular cell structure on $A \cup B$; i.e. $(A \cup B)^i = A^i \cup B^i$.

Now if $\mathscr{C}$ is a regular cell structure on S^n, it naturally induces a regular cell structure $c\mathscr{C}$ on $\mathbb{D}^{n+1}$ called the cone on $\mathscr{C}$ whose vertices are those of $\mathscr{C}$ together with $0 \in \mathbb{D}^{n+1}$ and whose open i-cells are the open i-cells of $\mathscr{C}$ together with the 0-deleted cones at 0 of the open $(i-1)$-cells of $\mathscr{C}$.

Example 2.24. See Figure 2.15.

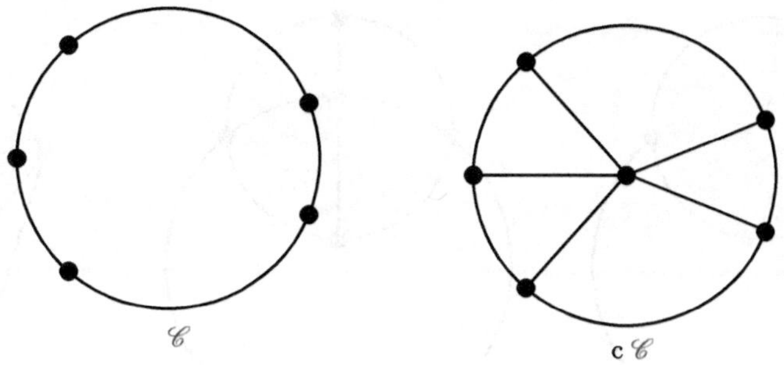

Figure 2.15

If $\mathscr{C}$ is a regular cell structure on X, then its barycentric subdivision $\mathrm{b}\mathscr{C}$ is a new regular cell structure on X defined *inductively* on the dimension of $\mathscr{C}$ (i.e. the maximal dimension of a cell in $\mathscr{C}$). If $\dim \mathscr{C} = 0$, then $\mathrm{b}\mathscr{C} = \mathscr{C}$. Suppose now that $\mathrm{b}\mathscr{C}$ is defined for all regular cell structures $\mathscr{C}$ with $\dim \mathscr{C} \leq n$. And now let $\mathscr{C}$ be a regular cell structure on X where $\dim \mathscr{C} = n+1$, set

$$\mathrm{b}\mathscr{C} = \mathrm{b}(\mathscr{C}_{X^n}) \cup \bigcup_{(n+1)\text{-cells } \sigma \in \mathscr{C}} \mathrm{c}(\mathrm{b}\mathscr{C}_{\partial\overline{\sigma}})$$

where $\partial\overline{\sigma} = \overline{\sigma} - \sigma$.

Remark 2.25. For this definition to work, we need that $\partial\overline{\sigma}$ is a subcomplex of $\mathscr{C}$; but this is a consequence of Invariance of Domain. Recall that this is the important property that if U and V are homeomorphic subsets of $\mathbb{R}^n$ and one of these subsets is open, then so is the other. Using this it is a good exercise to show that $\partial\overline{\sigma}$ is a subcomplex of $\mathscr{C}$.

An easier exercise is to prove the following important fact.

Proposition 2.26. *If $\mathscr{C}$ is a regular cell structure on X, then $\mathrm{b}\mathscr{C}$ is a triangulation of X; i.e. $\mathrm{b}\mathscr{C}$ is a simplicial complex.*

Example 2.27. See Figure 2.16.

Caveat 2.28. *In the handlebody decomposition*

$$W = \partial^- W \times [0,1] + h_1 + h_2 + \cdots + h_m$$

it is meant more precisely

$$W = (\cdots(((\partial^- W \times [0,1] + h_1) + h_2) + h_3) + \cdots) + h_m.$$

Example 2.29. Note that in the picture shown in Figure 2.17, $\partial^- W \times [0,1] + (h_1 + h_2)$ doesn't make sense in this example; while $(\partial^- W \times [0,1] + h_1) + h_2$ does.

The strategy for proving the h-Cobordism Theorem is:
Given any handlebody decomposition

$$W^n = \partial^- W \times [0,1] + h_1 + h_2 + \cdots + h_m$$

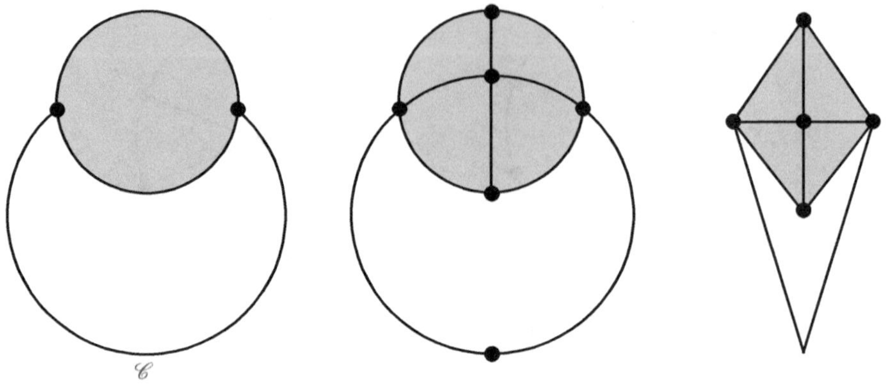

Figure 2.16

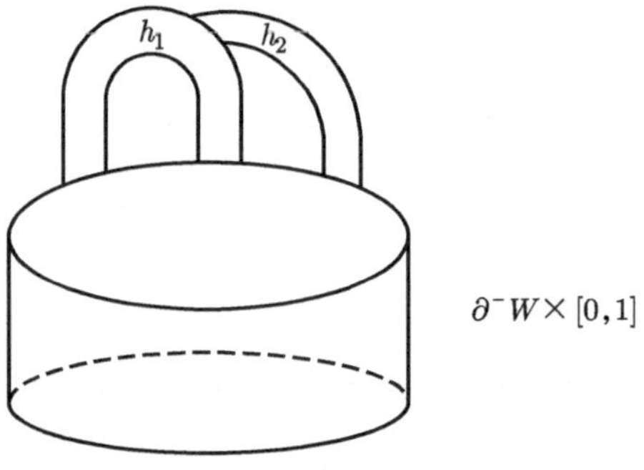

Figure 2.17

of an h-cobordism W^n ($\pi_1(W) = 1$, $n \geq 6$) and $m > 0$, we show how to produce a new one with *fewer* handles. Hence eventually we get one with no added handles; i.e.

$$W = \partial^- W \times [0, 1].$$

To show how to accomplish this reduction in the number of handles, we need to develop a calculus of handle moves.

2.4 Calculus of handle moves

Fact 1. If f and g are two *isotopic* attaching maps for the same handle h to the same manifold M (with non-empty boundary), then $M +_f h = M +_g h$.

This follows easily from the isotopy extension theorem as illustrated below where $F_t \colon \partial M \to \partial M$ is the ambient isotopy such that $F_1 \circ f = g$ (and $F_0 = \mathrm{id}_{\partial M}$). Let $F \colon \partial M \times [0, 1] \to \partial M \times [0, 1]$ be the self-diffeomorphism defined by $F(x, t) = (F_t(x), t)$. And we have identified $\partial M \times [0, 1]$ with a "closed" collar of

∂M in M; except now $\partial M \times 1$ is identified with ∂M instead of $\partial M \times 0$ (Figure 2.18).

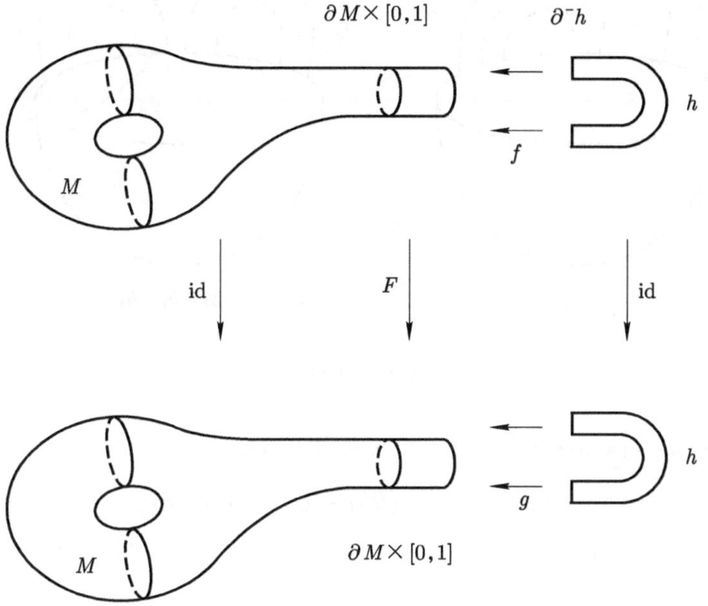

Figure 2.18

To proceed we need some more definitions.

Definition 2.30. *Each n-handle $h = \mathbb{D}^i \times \mathbb{D}^{n-i}$ of index i contains an* attaching *i-disc $e^i = \mathbb{D}^i \times 0$ and a* complementary *$(n-i)$-disc $c^{n-i} = 0 \times \mathbb{D}^{n-i}$ which intersect* transversally *in the point 0.*

Definition 2.31. *A pair of (smooth) submanifolds R^r, S^s of a manifold M^m meat* transversally *if either $R \cap S = \emptyset$ or, for each $x \in R \cap S$,*

$$T_xR + T_xS = T_xM$$

where the sum is not necessarily a direct sum. (Note for this to happen $r + s$ must be $\geq m$.)

The boundaries of these discs ∂e^i and ∂c^{n-i} are called the attaching and complementary spheres, respectively. Note that $\partial^- h = \partial e^i \times \mathbb{D}^{n-i}$ and $\partial^+ h = \mathbb{D}^i \times \partial c^{n-i}$.

Fact 2. Consider $M^n + h_1 + h_2 \colon = (M + h_1) + h_2$. Let $N^{n-1} = \partial(M + h_1)$ and note that the complementary sphere S for h_1 and the attaching sphere $\mathscr{S}$ for h_2 are submanifolds of N. If they are disjoint, then h_1 and h_2 can be attached simultaneously; denoted $M + h_1 + h_2 = M + (h_1 \sqcup h_2)$.

Example 2.32. See Figure 2.19.

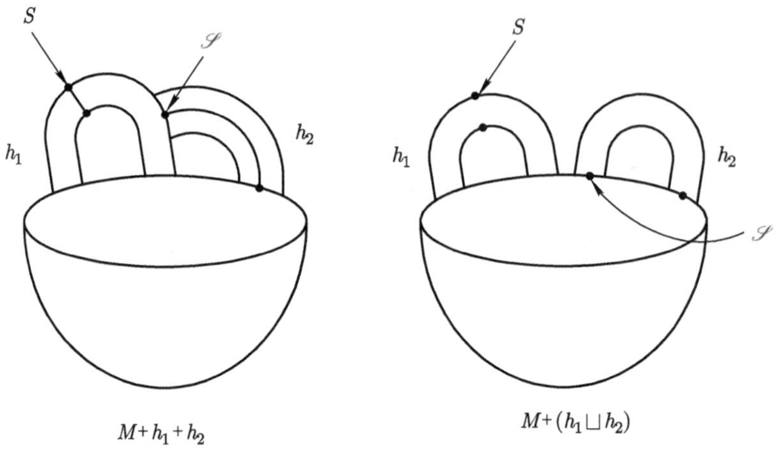

Figure 2.19

Fact 2 is a consequence of Fact 1. Note first that

$$\partial(M + h_1) = (\partial M - \partial^- h_1) \cup \partial^+ h_1$$

and by our assumption

$$\mathscr{S} \subset (\partial M - \partial^- h_1) \cup (\partial^+ h_1 - S).$$

Since $\partial^- h_2 = \mathscr{S} \times \mathbb{D}^{n-i}$, where $i = I(h_2)$, we can isotope the attaching map $f\colon \partial^- h_2 \to \partial(M + h_1)$, by "squeezing" $\mathbb{D}^{n-i}$ towards $0 \in \mathbb{D}^{n-i}$, so that

$$f(\partial^- h_2) \subset (\partial M - \partial^- h_1) \cup (\partial^+ h_1 - S);$$

i.e. $f(\partial^- h_2) \cap S = \emptyset$.

Now using that

$$\partial^+ h_1 - S = (\mathbb{D}^j - 0) \times S$$

where $j = I(h_1)$, we can isotope $f(\partial^- h_2)$ off of $(\mathbb{D}^j - 0) \times S$ by pushing it towards $\partial((\mathbb{D}^j - 0) \times S) = S^{j-1} \times S \subset \partial M$. Now Fact 1 shows that h_1 and h_2 can be attached simultaneously; i.e.

$$M + h_1 + h_2 = M + (h_1 \sqcup h_2).$$

Remark 2.33. It is a consequence of Fact 2 (or by examining the proof of Theorem 2.15 using Whitehead's Triangulation Theorem) that any cobordism W^n contains a filtration by compact codimension-0 submanifolds

$$\partial^- W \times [0,1] = W_{-1} \subseteq W_0 \subseteq W_1 \subseteq \cdots \subseteq W_{n-1} \subseteq W_n = W$$

such that

$$W_i = W_{i-1} + (h_1^i \sqcup h_2^i \sqcup \cdots \sqcup h_{m_i}^i)$$

where $i \geq 0$, $I(h_j^i) = i$ and m_i is the number of i-handles in this handlebody decomposition. Such a decomposition we call *standard*.

To see this one uses the following basic result.

Thom Transversality Theorem. *Let K and L be smooth submanifolds of a smooth manifold M, then we can "wiggle" (i.e. by a short isotopy) K slightly to an isotopic new embedding K' such K' is transverse to L.*

(See Milnor-Stasheff [17] for more details.)

Now consider $M^n + h_1 + h_2$ as in Fact 2 and wiggle $\mathscr{S}$ inside N^{n-1} so that it is transverse to S. And note that if $I(h_2) \leq I(h_1)$, then $\dim \mathscr{S} + \dim S < n - 1 = \dim N$. Hence $\mathscr{S} \cap S = \emptyset$ and $M^n + h_1 + h_2 = M^n + (h_1 \sqcup h_2)$ by Fact 2. This is the key step in verifying the above remark.

We can use the attaching and complementary spheres to define n incidence matrices (with integral entries) $D^1, D^2, \cdots, D^n$ provided our handle decomposition is not only standard, but *t-standard* (transverse standard), which is defined as follows. Let

$$N_i = \partial^+ W_i = \partial W_i - \partial^- W_i$$

and $\mathscr{S}_1, \mathscr{S}_2, \cdots, \mathscr{S}_{m_{i+1}}$ be the attaching spheres for the handles of index $i+1$. They form a disjoint family of submanifolds of N_i^{n-1}; each diffeomorphic to S^i. Let $S_1, S_2, \cdots, S_{m_i}$ be the complementary spheres for the handles of index i. They form a second disjoint family of submanifolds of N_i^{n-1}; each diffeomorphic to $S^{n-(i+1)}$. By the Thom Transversality Theorem we may assume that each S_j intersects each $\mathscr{S}_k$ transversally, and hence in a finite number of points since $\dim S_j + \dim \mathscr{S}_k = \dim N_i = n - 1$. When this happens the handle decomposition is called t-standard. For t-standard decompositions each intersection point $x \in S_j \cap \mathscr{S}_k \subset N_i$ is equipped with a sign $\sigma(x) = \pm 1$ determined as follows. The two spheres S_j and $\mathscr{S}_k$ are naturally oriented and hence determine an orientation for $T_x N_i$. But W can also be oriented since $\pi_1(W) = 0$; so fix an orientation for W. It determines an orientation for each codimension-0 submanifold W_i^n and hence for $\partial^+ W_i^n = N_i^{n-1}$. If the two orientations for $T_x N_i^{n-1}$ agree, then assign $+1$ to x; if not, assign -1. Then the (j,k)-th entry $D_{j,k}^{i+1}$ equals the sum over the points in $S_j \cap \mathscr{S}_k$ of these thus assigned orientation numbers; i.e.

$$D_{j,k}^{i+1} = \sum_{x \in S_j \cap \mathscr{S}_k} \sigma(x).$$

Example 2.34. See Figure 2.20.

The incidence matrices D^i define a chain complex

$$C \colon C_n \overset{D^n}{\to} C_{n-1} \overset{D^{n-1}}{\to} C_{n-2} \to \cdots \overset{D^2}{\to} C_1 \overset{D^1}{\to} C_0$$

where the bases for the chain groups C_i are the handles of index i in the given t-standard handlebody decomposition of W, such that $H_i(C) = H_i(W, \partial W)$ for all i. This is geometrically what should be true. And a complete proof can be derived from §9 of Milnor's long survey paper [15].

Dual handlebody decomposition and Lefschetz duality

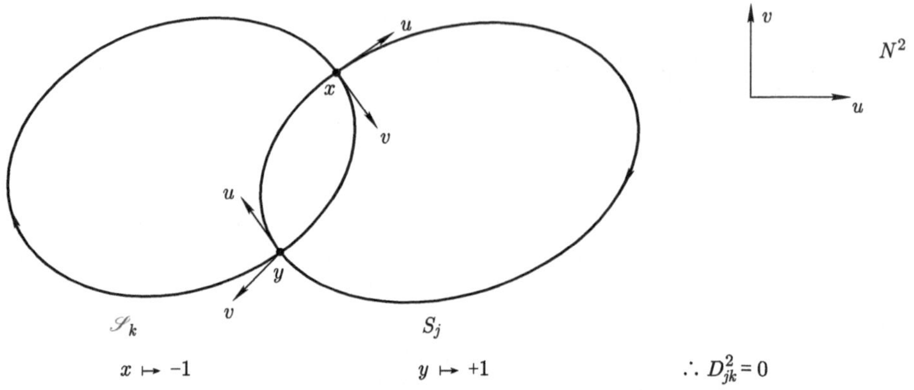

Figure 2.20

Fixing a t-standard handlebody decomposition for the cobordism W^n from $\partial^- W$ to $\partial^+ W$ as above, it determines a "dual" t-standard handlebody decomposition for "W turned upside down" which we denote by $-W$. Here $\partial^-(-W) = \partial^+ W$ and $\partial^+(-W) = \partial^- W$. See Figure 2.21.

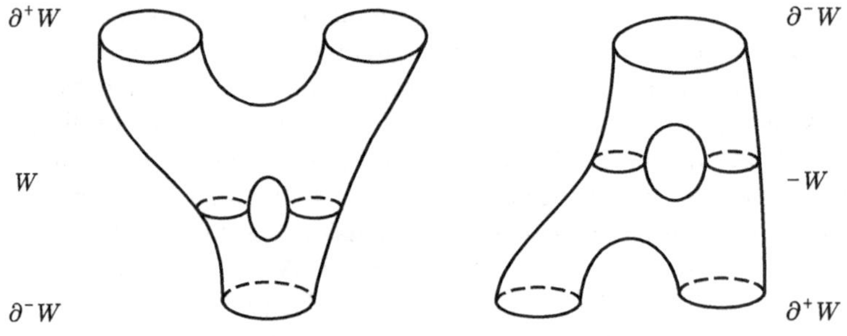

Figure 2.21

This is a consequence of the obvious fact that the n-dimensional handle $h = \mathbb{D}^i \times \mathbb{D}^{n-i}$ of index i when "turned upside down" becomes the handle $-h = \mathbb{D}^{n-i} \times \mathbb{D}^i$ of index $n-i$. Also notice that the attaching sphere $\mathscr{S}$ of h is the complementary sphere of $-h$, and the complementary sphere S of h is the attaching sphere of $-h$. Therefore

$$\partial^- W \times [0,1] + h_1 + \cdots + h_m = \partial^+ M \times [0,1] + (-h_m) + (-h_{m-1}) + \cdots + (-h_1).$$

The family of codimension-1 submanifolds of $-W$, $\{\mathscr{N}_i \mid i = 0, 1, \cdots, n-1\}$, which separate the index i and index $(i+1)$ handles is the same as the family $\{N_i \mid i = 0, 1, \cdots, n-1\}$ for W; but dually renumbered; i.e.

$$\mathscr{N}_i = N_{(n-1)-i}.$$

Furthermore if $S_1, S_2, \cdots, S_{m_i}$ and $\mathscr{S}_1, \mathscr{S}_2, \cdots, \mathscr{S}_{m_{i+1}}$ are the complementary and attaching spheres for the i and $(i+1)$ handles (respectively) of the given decomposition, then $\mathscr{S}_1, \mathscr{S}_2, \cdots, \mathscr{S}_{m_{i+1}}$ and $S_1, S_2, \cdots, S_{m_i}$ are the complementary and attaching spheres for the $((n-1)-i)$ and $(n-i)$ handles (respectively) of the dual decomposition.

We now proceed to calculate the incidence matrices $\mathscr{D}^1, \mathscr{D}^2, \cdots, \mathscr{D}^n$ for the dual decomposition of $-W$. Let $x \in \mathscr{S}_j \cap S_k$ and $\sigma(x) = \pm 1$ be the sign given to x in the given handle decomposition of W; while $\overline{\sigma}(x)$ is the sign given to x in the dual decomposition of $-W$. Then

$$\overline{\sigma}(x) = (-1)^{i(n-(i+1))}\sigma(x)$$

which is seen by noticing that the determinant of the change of basis matrix relating the following two bases of $\mathbb{R}^{n-1}$

$$u_1, u_2, \cdots, u_i; v_1, v_2, \cdots, v_{n-(i+1)}$$

and

$$v_1, v_2, \cdots, v_{n-(i+1)}; u_1, u_2, \cdots, u_i$$

is $(-1)^{i(n-(i+1))}$. (For this to be the correct formula, we give $-W$ the negative of the (global) orientation for W.) Hence

$$\mathscr{D}^{n-i}_{k,j} = (-1)^{i(n-(i+1))} D^{i+1}_{j,k};$$

i.e. up to multiplication by ± 1 $\mathscr{D}^i$ is $(D^{(n+1)-i})^*$; i.e. the transpose of $D^{(n+1)-i}$. Therefore $H_i(W, \partial^+ W)$ is $\ker(D^{n-i})^* / \operatorname{im}(D^{(n+1)-i})^*$, which is, by definition, the $(n-i)$-th cohomology group of the chain complex

$$C\colon C_n \overset{D^n}{\to} C_{n-1} \overset{D^{n-1}}{\to} C_{n-2} \to \cdots \overset{D^2}{\to} C_1 \overset{D^1}{\to} C_0.$$

Thus we have proven the following version of Lefschetz duality.

Theorem 2.35. *Let W^n be a smooth (compact and simply-connected) cobordism from $\partial^- W$ to $\partial^+ W$, then*

$$H_i(W, \partial^+ W) \cong H^{n-i}(W, \partial^- W)$$

for all i.

Remark 2.36. We so far only used the simply-connected assumption to conclude that W^n is orientable. So in the above result, we can replace "$\pi_1(W) = 0$" by "W is orientable". Ponicaré duality and the standard form of Lefschetz duality can be proven in an analogous manner provided the manifold is smooth and orientable.

Fact 3. (geometric cancellation) $M + h_1 + h_2 = M$ if the attaching sphere $\mathscr{S}$ for h_2 intersects the complementary sphere S for h_1 transversally in a *single* point.

Note that this can only happen when $I(h_2) = I(h_1) + 1$. The proof of this result is an exercise in the differential topology techniques developed so far in this course. The following two examples illustrate how to prove Fact 3.

Example 2.37. In the example shown in Figure 2.22, $I(h_1) = 0$, $I(h_2) = 1$ and $n = 3$. Here $S = S^2 = \partial h_1$ and $\mathscr{S} = S^0 = 2$ points.

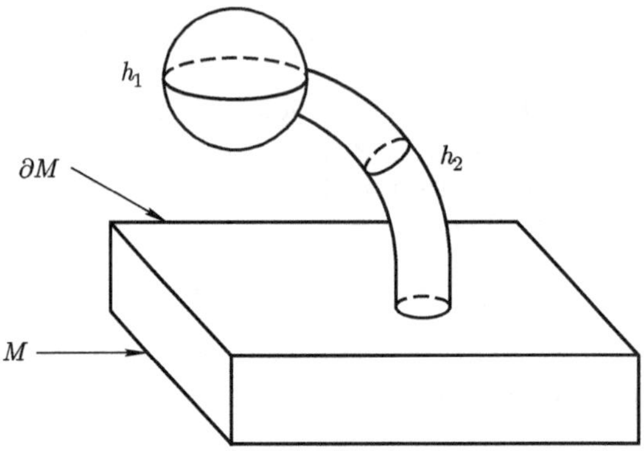

Figure 2.22

Note that $h_1 \cup h_2$ is diffeomorphic to $\mathbb{D}^3$ and $(h_1 \cup h_2) \cap \partial M$ is diffeomorphic to $\mathbb{D}^2$. Hence $M + h_1 + h_2$ is diffeomorphic to M with $h_2 = \mathbb{D}^2 \times [0, 1]$ attached via an embedding of $\mathbb{D}^2 \times 0$ into ∂M (Figure 2.23).

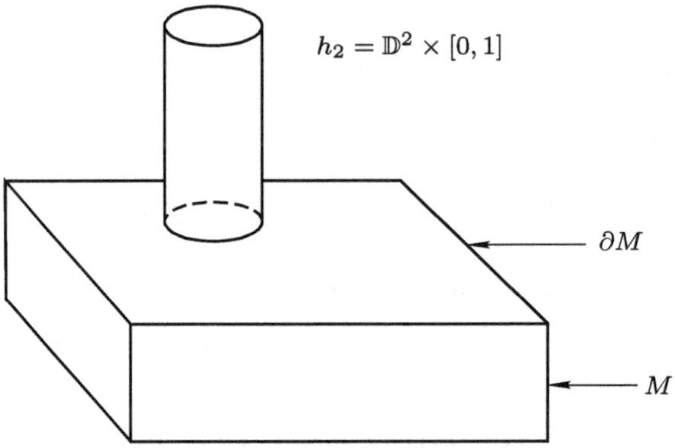

Figure 2.23

And the result is easily seen to be diffeomorphic to M.

Example 2.38. In Figure 2.24 $I(h_1) = 1$, $I(h_2) = 2$ and $n = 3$.

Here S and $\mathscr{S}$ are both diffeomorphic to S^1 and intersect in a single point as shown. And after an isotopy one can arrange that $\partial^- h_2 = \mathscr{S} \times \mathbb{D}^1$ intersects a small neighborhood $(-\epsilon, \epsilon) \times \mathbb{D}^2$ of the complementary disc $0 \times \mathbb{D}^2$ in h_1 canonically in $(-\epsilon, \epsilon) \times (-\epsilon, \epsilon)$. Then $M^n + h_1 + h_2 = ((M^n \cup A) \cup B) \cup C$ where

$$A = h_1 - (-\epsilon, \epsilon) \times \mathbb{D}^2,$$
$$B = h_2,$$
$$C = [-\epsilon, \epsilon] \times \mathbb{D}^2.$$

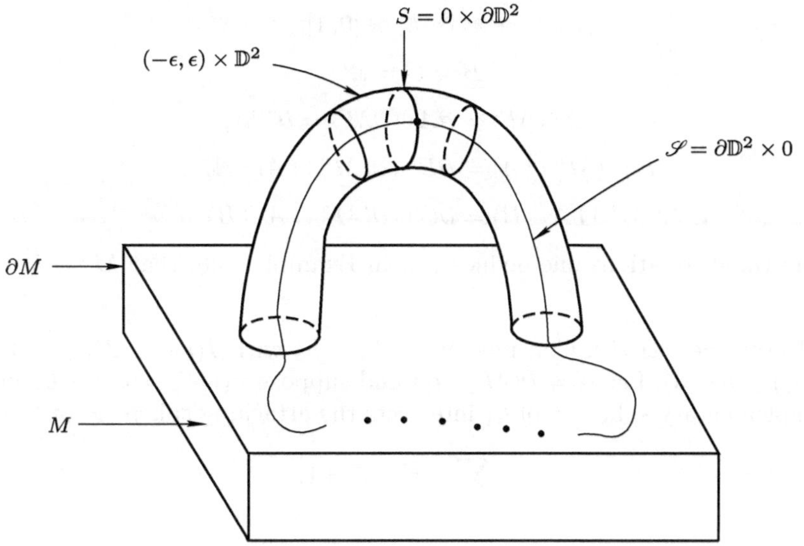

Figure 2.24

Note that B and C are each diffeomorphic to $\mathbb{D}^3$. And that

$$(M^n \cup A) \cap B = \partial(M^n \cup A) \cap \partial B = \mathbb{D}^{n-1},$$

$$(M^n \cup A \cup B) \cap C = \partial(M^n \cup A \cup B) \cap \partial C = \mathbb{D}^{n-1}.$$

Also $A = \partial^- A \times [0,1]$ where $\partial^- A = A \cap M = \partial A \cap \partial M$. And this analysis implies that

$$M^n \cup A = M^n,$$
$$(M^n \cup A) \cup B = M^n \cup A,$$
$$(M^n \cup A \cup B) \cup C = M^n \cup A \cup B.$$

Consequently $M^n + h_1 + h_2 = M^n$.

The general situation of Fact 3 follows from the same pattern of proof.

General proof of Fact 3. Here $h_1 = \mathbb{D}^i \times \mathbb{D}^{n-i}$ and $h_2 = \mathbb{D}^{i+1} \times \mathbb{D}^{(n-1)-i}$. We again define codimension-0 compact submanifolds A, B and C of $M^n + h_1 + h_2$ such that $M^n + h_1 + h_2 = ((M^n \cup A) \cup B) \cup C$ by

$$A = \text{Closure}(h_1 - \epsilon\mathbb{D}^i \times \mathbb{D}^{n-i}),$$
$$B = h_2,$$
$$C = \epsilon\mathbb{D}^i \times \mathbb{D}^{n-i},$$

where $\epsilon\mathbb{D}^i$ is the concentric i-dimensional ball of radius ϵ. And after an isotopy we can arrange that $\partial^- h_2 = \mathscr{S} \times \mathbb{D}^{(n-1)-i}$ intersects C in $\epsilon\mathbb{D}^i \times \epsilon\mathbb{D}^{(n-1)-i}$. Then one easily checks the following

$$\begin{aligned}
A &= \partial^- h_1 \times [0,1], \\
B &= C = \mathbb{D}^n, \\
A \cap M^n &= \partial A \cap \partial M^n = \partial^- h_1, \\
B \cap (M^n \cup A) &= \partial B \cap \partial(M^n \cup A) = \mathbb{D}^{n-1}, \\
C \cap (M^n \cup A \cup B) &= \partial C \cap \partial(M^n \cup A \cup B) = \mathbb{D}^{n-1}.
\end{aligned}$$

From these equations one deduces, as in Example 2.38, that $M^n + h_1 + h_2 = M^n$. □

Fact 4 (Whitney Trick). Consider $M^n + h_1 + h_2$ with $I(h_2) = I(h_1) + 1$. (Also $2 \leq I(h_1) \leq n-3$). Let $N = \partial(M + h_1)$ and suppose $\pi_1(N) = 0$, $n \geq 6$, and that the complementary sphere S of h_1 intersects the attaching sphere $\mathscr{S}$ of h_2 so that

$$\sum_{x \in \mathscr{S} \cap S} \sigma(x) = \pm 1.$$

Then the attaching map $f\colon \partial^- h_2 \to N^{n-1}$ can be isotoped so that the new attaching sphere $\mathscr{S}'$ intersects S transversally in exactly one point.

Combining Facts 1, 3 and 4 we get the following consequence.

Fact 4'. (algebraic cancellation) Under the assumption of Fact 4 (i.e. $\pi_1(N) = 0$, $n \geq 6$ and $\sum_{x \in \mathscr{S} \cap S} \sigma(x) = \pm 1$) we have $M^n + h_1 + h_2 = M^n$.

The proof of Fact 4 uses deep work of Whitney to isotope $\mathscr{S}$ so as to reduce the number of points in the intersection $S \cap \mathscr{S}$ by two at a time. Crucial use of the two hypotheses

$$\pi_1(N) = 0 \text{ and } n \geq 6$$

is made to do this. The following is a brief sketch of the proof of Whitney Trick.

First pick a pair of points x, $y \in S \cap \mathscr{S}$ such that $\sigma(x) = 1$ and $\sigma(y) = -1$. Connect x to y by a pair of smooth embedded arcs α and γ where $\alpha\colon [0,1] \to S$ and $\gamma\colon [0,1] \to \mathscr{S}$ with $\alpha(0) = x = \gamma(0)$, $\alpha(1) = y = \gamma(1)$ and $\alpha((0,1)) \cap \gamma((0,1)) = \emptyset$. The concatenation $\gamma^{-1} * \alpha$ is an embedded loop in the simply-connected manifold N^{n-1}; hence it spans a disc $\mathbb{D}^2$ which can be chosen to be embedded if $n - 1 \geq 5$ (i.e. $n \geq 6$) by the Whitney Embedding Theorem. Now isotope $\mathscr{S}$ in a neighborhood of α off of S by pushing along this disc.

Example 2.39. In the example shown in Figure 2.25, $n = 4$, $I(h_1) = 1$ and $I(h_2) = 2$. Then $S = S^2$, $\mathscr{S} = S^1$ and $\dim N = 3$. (Of course n is *not* bigger than 5 and $\pi_1(N) \neq 1$ in this example. But we have chosen our example so that the possible obstructions vanish in this case.)

Fact 5. (handle addition) Given a t-standard handle decomposition of a compact smooth oriented cobordism W^n with incidence matrices D^i where $i = 1, 2, \cdots, n$. We let m_i be the number of index i-handles in this decomposition. And focus on an index i, where $2 \leq i \leq n-2$. If we fix a pair of distinct integers j and k where

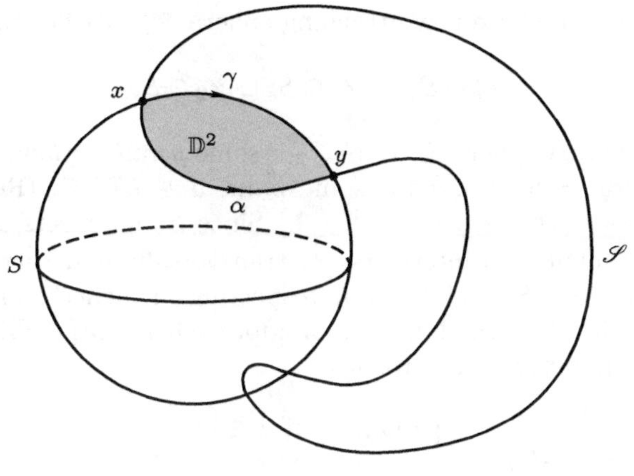

Figure 2.25

$1 \leq j, k \leq m_i$, then we can construct a new t-standard handle decomposition for W^n with the same number of handles of each index but whose new i-th incidence matrix $\mathscr{D}^i$ is obtained from D^i by either adding (or subtracting) the k-th column on D^i to (from) its j-th column.

Addendum 2.40. *Let us again assume the first three sentences of Fact* 5, *and let* $\mathscr{D}^i$ *be any matrix obtainable from* D^i *by a finite sequence of (integral) elementary column operations. Then there exists a new t-standard handle decomposition for* W^n *with the same number of handles of each index* j*; but such that* $\mathscr{D}^i$ *is its* i*-th incidence matrix.*

Proof of Fact 5. We make use of the notation from Remark 2.33; i.e. there is a filtration

$$\partial^- W \times [0,1] = W_{-1} \subseteq W_0 \subseteq W_1 \subseteq \cdots \subseteq W_{n-1} \subseteq W_n = W$$

such that

$$W_s = W_{s-1} + (h_1^s \sqcup h_2^s \sqcup \cdots \sqcup h_{m_s}^s)$$

where $I(h_t^s) = s$ and $N_s = \partial^+ W_s$. To simplify our notation, we now assume that $j = 1$ and $k = 2$ which causes no loss in generality. Also we abbreviate h_t^i by h_t and N_i by N. As before $\mathscr{S}_1, \mathscr{S}_2, \cdots, \mathscr{S}_{m_i}$ and $S_1, S_2, \cdots, S_{m_{i-1}}$ denote the attaching and complementary spheres (respectively) for the index i and $i - 1$ handles. Recall these are spheres embedded in N^{n-1}. To get the t-standard handlebody decomposition posited in Fact 5 from the given one, we isotope the attaching embedding for $\partial^- h_1$ inside

$$N - (\partial^- h_2 \sqcup \partial^- h_3 \sqcup \cdots \sqcup \partial^- h_{m_i}) \cup \partial^+ h_2$$

to a new embedding for $\partial^- h_1$ which again lies in

$$N - (\partial^- h_2 \sqcup \partial^- h_3 \sqcup \cdots \sqcup \partial^- h_{m_i})$$

(i.e. misses $\partial^+ h_2$) and whose new attaching sphere $\mathscr{S}_1'$ satisfies the following:

$$\mathscr{S}_1' \cap S_s = \mathscr{S}_1 \cap S_s \sqcup \hat{\mathscr{S}}_2 \cap S_s$$

for each complementary sphere S_s. Here $\hat{\mathscr{S}}_2$ is some *parallel sphere* to the attaching sphere $\mathscr{S}_2$; i.e. $\hat{\mathscr{S}}_2 = S^{i-1} \times a$ for some point $a \in \partial\mathbb{D}^{n-i}$. (Recall that $h_2 = \mathbb{D}^i \times \mathbb{D}^{n-i}$ and $\mathscr{S}_2 = \partial\mathbb{D}^i \times 0 = S^{i-1} \times 0$.) Since S_s intersects $\mathscr{S}_2$ transversally we may assume that it also intersects $\hat{\mathscr{S}}_2$ transversally and there is a canonical bijection between $\mathscr{S}_2 \cap S_s$ and $\hat{\mathscr{S}}_2 \cap S_s$ respecting incidence numbers. Hence we identify $\hat{\mathscr{S}}_2 \cap S_s$ with $\mathscr{S}_2 \cap S_s$. Furthermore, for each $x \in \mathscr{S}_1' \cap S_s$, we can arrange that its incidence number $\sigma'(x)$ satisfies

$$\sigma'(x) = \begin{cases} \sigma(x), & \text{if } x \in \mathscr{S}_1, \\ \pm\sigma(x), & \text{if } x \in \hat{\mathscr{S}}_2 \equiv \mathscr{S}_2. \end{cases}$$

The $\pm$ sign in this formula depends on whether in Fact 5 we desire that the j-th column of $\mathscr{D}^i$ be the j-th column of D^i *plus* the k-th column of D^i or the j-th column *minus* the k-th column.

The desired isotopy of the given embedding of $\partial^- h_1$ is constructed as follows. First choose a smooth embedded arc $\alpha\colon [0,1] \to N^{n-1}$ which intersects the union of the attaching spheres $\mathscr{S}_s$ in exactly two points: $\alpha(0) \in \mathscr{S}_1$ and $\alpha(1) \in \mathscr{S}_2$. This is possible since $n - i \geq 2$ (Figure 2.26).

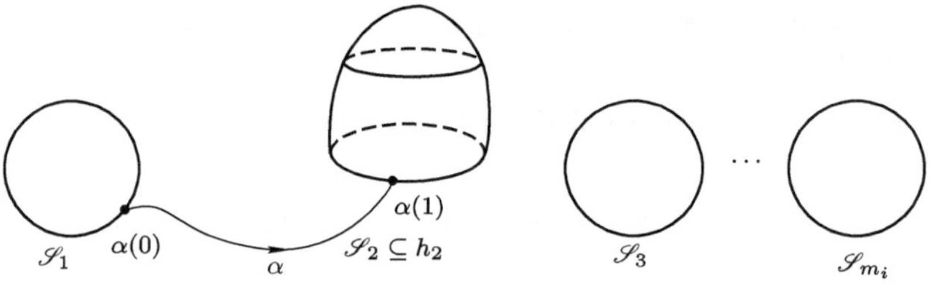

Figure 2.26

Then push a small ball $\mathbb{D}^{i-1}$ centered at $\alpha(0) \in \mathscr{S}_1 = S^{i-1}$ along α and over the handle h_2 so that the new attaching sphere $\mathscr{S}_1'$ (for h_1) is either the connected sum $\mathscr{S}_1 \sharp \hat{\mathscr{S}}_2$ (if we want $\sigma'(x) = \sigma(x)$ for $x \in \hat{\mathscr{S}}_2 \cap S_s$) or $\mathscr{S}_1 \sharp - \hat{\mathscr{S}}_2$ (if we want $\sigma'(x) = -\sigma(x)$).

Example 2.41. $i = 2$, $n = 4$ and $N^{n-1} = S^3$ (Figure 2.27).

To be able to accomplish this for the case $\sigma'(x) = -\sigma(x)$ (as illustrated above) we again use that $n - i \geq 2$. And this isotopy of $\mathscr{S}_1$ is easily extended to all of $\partial^- h_1$ using that $\partial^- h_1 = \mathscr{S}_1 \times \mathbb{D}^{n-i}$.

Finally observe that

$$W_i = W_{i-1} + (h_1 \sqcup h_2 \sqcup \cdots \sqcup h_{m_i}) = (W_{i-1} + (h_2 \sqcup h_3 \sqcup \cdots \sqcup h_{m_i})) + h_1$$

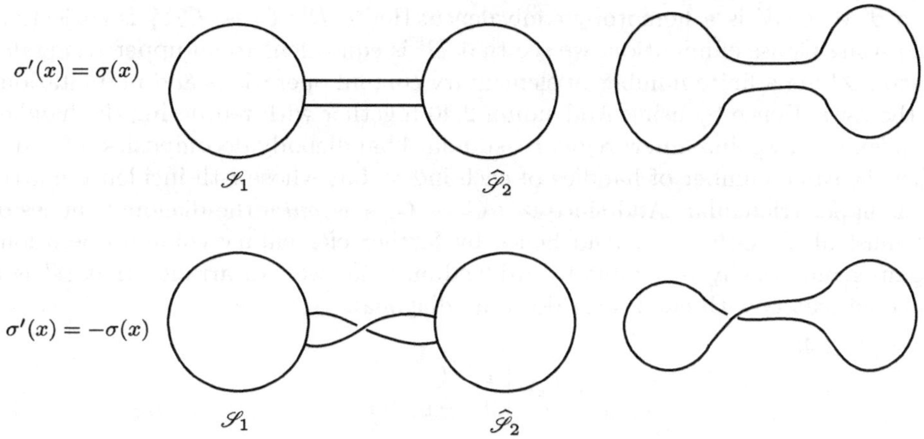

Figure 2.27

and that, by Fact 1, isotopic embeddings of $\partial^- h_1$ in $\partial^+(W_{i-1}+(h_2 \sqcup h_3 \sqcup \cdots \sqcup h_{m_i}))$ give diffeomorphic cobordisms. □

2.5 Proof of the *h*-Cobordism Theorem

We now prove Smale's h-Cobordism Theorem modulo the following (reasonable) assumption which will be demonstrated shortly afterwards.

Proposition 2.42. *Every smooth h-cobordism W^n $(n \geq 6)$ has a t-standard handlebody decomposition which does not contain any handle of indices 0, 1, $n-1$ and n. Such a decomposition is called* regular.

Proof of the h-Cobordism Theorem. Let

$$W^n = \partial^- W \times [0,1] + h_1 + h_2 + \cdots + h_m$$

be a regular t-standard handlebody decomposition of W^n; i.e. with *no* handles of indices 0, 1, $n-1$ or n. Recall that our strategy is to show that if $m > 0$, then there exists another regular t-standard decomposition with fewer handles. If we can do this, then there obviously must be a decomposition with *no* handles; hence $W^n = \partial^- W^n \times [0,1]$ thus proving the h-Cobordism Theorem.

Let $i-1$ be the smallest index of a handle in the given decomposition. Hence the first non-zero group in the chain complex C (described in §2.4) which computes $H_*(W, \partial^- W)$ is C_{i-1}; i.e. $C_j = 0$ for $j = 0, 1, \cdots, i-2$. Therefore

$$H_{i-1} = C_{i-1}/\operatorname{im}(D^i)$$

where $D^i \colon C_i \to C_{i-1}$ is the incidence matrix. But $H_{i-1}(W, \partial^- W) = 0$ which follows from the long exact sequence for the pair $(W, \partial^- W)$ since the inclusion map $\partial^- W \subset W$ is a homotopy equivalence. Hence $D^i \colon C_i \twoheadrightarrow C_{i-1}$ is surjective. And using Gauss elimination, we see that D^i is equivalent to an upper triangular matrix $\mathscr{D}^i$ via a finite number of elementary column operations and permutations of the rows. Hence by using Addendum 2.40 together with reordering the handles of index $i-1$, we find a new regular t-standard handlebody decomposition for W^n with the same number of handles of each index; but whose i-th incidence matrix $\mathscr{D}^i$ is upper triangular. And since $\mathscr{D}^i \colon C_i \twoheadrightarrow C_{i-1}$ is *onto*, the diagonal entries of $\mathscr{D}^i$ must all be either ± 1. And hence by further elementary column operations (again geometrically realizable by Addendum 2.40) we can arrange that $\mathscr{D}^i$ is a diagonal matrix with entries ± 1 down its diagonal.

Example 2.43.

$$\mathscr{D}^i = \begin{bmatrix} 1 & 0 & 0 \\ 0 & -1 & 0 \end{bmatrix}.$$

Recall the filtration $W_{-1} \subset W_0 \subset \cdots \subset W_{i-1} \subset W_i \subset \cdots \subset W_n = W$. By our assumption

$$\partial^- W \times [0,1] = W_{-1} = W_{i-2},$$
$$W_{i-1} = W_{i-2} + (h_1^{i-1} + \cdots + h_{m_{i-1}}^{i-1})$$

and

$$W_i = W_{i-1} + (h_1^i + \cdots + h_{m_i}^i);$$

i.e. W_{i-1} is W_{i-2} with all the handles of index $i-1$ simultaneously attached, and likewise for W_i. Recall the closed codimension-1 submanifold N_{i-1} of W is $\partial^+ W_{i-1}$ and contains the complementary spheres $S_1, \cdots, S_{m_{i-1}}$ for the $(i-1)$-handles, and the attaching spheres $\mathscr{S}_1, \cdots, \mathscr{S}_{m_i}$ for the i-handles whose transverse intersection determine the diagonal matrix $\mathscr{D}^i$.

Example 2.44. In the example shown in Figure 2.28, the matrix $\mathscr{D}^i$ is as above,

$$\mathscr{D}^i = \begin{bmatrix} 1 & 0 & 0 \\ 0 & -1 & 0 \end{bmatrix}.$$

Letting $\hat{h}_1 = h_1^{i-1}$ and $\hat{W}_{i-1} = W_{i-2} + (h_2^{i-1} \sqcup \cdots \sqcup h_{m_{i-1}}^{i-1})$ we have that

$$W_{i-1} = \hat{W}_{i-1} + \hat{h}_1.$$

Similarly letting $\hat{h}_2 = h_1^i$ we have

$$W_i = \hat{W}_i + (h_2^i \sqcup \cdots \sqcup h_{m_i}^i)$$

where $\hat{W}_i = W_{i-1} + \hat{h}_2$. Notice that we have thus refined the given filtration so that

$$W_{i-2} \subset \hat{W}_{i-1} \subset W_{i-1} \subset \hat{W}_i \subset W_i.$$

Now note that

$$\hat{W}_i = \hat{W}_{i-1} + \hat{h}_1 + \hat{h}_2$$

and

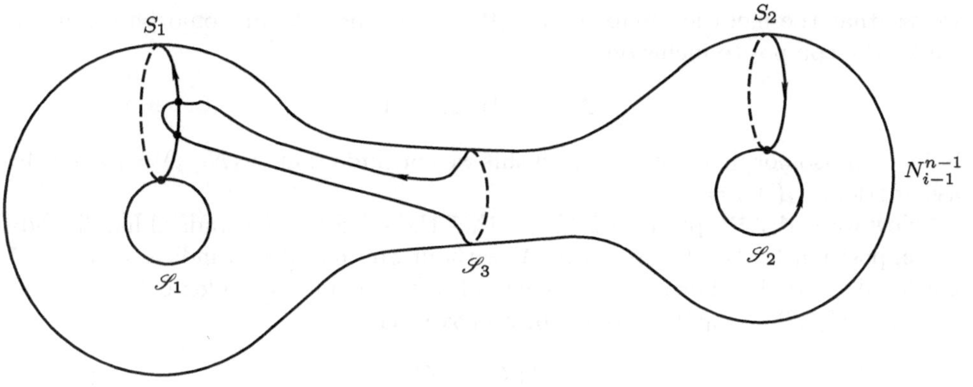

Figure 2.28

$$W_{i-1} = \hat{W}_{i-1} + \hat{h}_1,$$

where $\partial(\hat{W}_{i-1} + \hat{h}_1) = N_{i-1}$. Now apply Fact 4' with

$$M^n = \hat{W}_{i-1},\ h_1 = \hat{h}_1,\ h_2 = \hat{h}_2, \text{ and } N = N_{i-1}.$$

For this we need that $\pi_1(N_{i-1}) = 1$ which is proven below. Also $2 \leq i - 1 = I(h_1) \leq n - 3$ since the given handlebody decomposition is regular. Hence we conclude that $\hat{W}_i = \hat{W}_{i-1}$; i.e. we can cancel handles h_1^{i-1} and h_1^i in the given handle decomposition resulting a new regular t-standard decomposition for W with *fewer handles.* This completes the proof of the h-Cobordism Theorem modulo showing that $\pi_1(N_{i-1}) = 1$ and verifying Proposition 2.42. □

Our demonstration that $\pi_1(N_{i-1}) = 1$ depends on the following well-known result.

Lemma 2.45. *Let $A = B \cup_\alpha \mathbb{D}^s$ where $\alpha \colon S^{s-1} \to B$; i.e. A is obtained from B by attaching an s-dimensional cell to B via α. If $s \geq 3$ and A is path connected, then the inclusion $B \subset A$ induces an isomorphism of fundamental groups.*

Proof. This results immediately from applying Van Kampen Theorem to the decomposition

$$A = (A - 0) \cup \mathrm{Int}(\mathbb{D}^s)$$

where 0 is the center of $\mathbb{D}^s$. □

Now observe that $W = W_{i-1} + k_1 + \cdots + k_s$ where $\{k_j \mid j = 1, 2, \cdots, s\}$ are the handles of the given decomposition of W^n having index $\geq i$. Using this, it can be shown that W is homotopy equivalent to W_{i-1} with a finite sequence of cells of dimension ≥ 3 attached, since $i \geq 3$. Now inductively applying the above lemma to this cell structure, we conclude that the inclusion map $W_{i-1} \subseteq W$ induces an isomorphism on π_1. Next recall that W_{i-1} is a cobordism from $\partial^- W$ to N_{i-1} having only handles of index $i - 1$. Hence $-W_{i-1}$ has a dual decomposition from N_{i-1} to $\partial^- W$ having only handles of index $(n - i) + 1 \geq 3$. Arguing as before, we see that the inclusion map $N_{i-1} \subseteq W_{i-1}$ also induces an isomorphism on π_1. Hence the composite inclusion

$$N_{i-1} \subseteq W_{i-1} \subseteq W$$

induces an isomorphism on π_1. From this we conclude that $\pi_1(N_{i-1}) = 1$ since, by assumption, $\pi_1(W) = 1$.

Now we verify Proposition 2.42; i.e. that there exists a t-standard handlebody decomposition for the h-cobordism W without any handles of indices 0, 1, $n-1$ and n. We start by showing how to get rid of the handles of index 0.

Since $H_0(C) = H_0(W, \partial^- W) = 0$, we have that

$$D^1 \colon C_1 \twoheadrightarrow C_0$$

is onto. Assuming that $C_0 \neq 0$; i.e. there is at least one index 0 handle in the given decomposition, then an algebraic-combinatorial argument shows there exists a pair of handles h_0, h_1 with $I(h_0) = 0$, $I(h_1) = 1$, attaching sphere $\mathscr{S} = S^0$ for h_1, complementary sphere $S = S^{n-1} = \partial^+ h_0$ for h_0, and such that

$$S \cap \mathscr{S} = \{pt\} \text{ and } \mathscr{S} \cap (\partial^- W \times 1) = \{pt\}.$$

Hence by Fact 4, $\partial^- W \times [0,1] + h_0 + h_1 = \partial^- W \times [0,1]$ and thus we get a new t-standard decomposition of W with one fewer handle of both index 0 and index 1 by canceling these two handles, but the same number of index i handles for $i > 1$. Continuing in this way we get a new t-standard decomposition, denoted $*$, containing *no* index 0 handles, but the same number of index i handles for $i > 1$. Hence its *dual decomposition* $-*$ contains *no* index n handles.

Then we apply the above reduction procedure to $-*$ eliminating all of its index 0 handles without introducing, in particular, *any* index n handles. Denote this resulting handle decomposition of $-W$ by $\square$. Note that $\square$ contains *no* handles of either index 0 or index n. Hence its dual $-\square$ is a t-standard decomposition for W also containing *no* handles of either index 0 or index n.

Trading Handles

To get rid of the index 1 handles, and by applying it to the dual decomposition as above, the index $(n-1)$ handles also, we introduce an important new procedure called *handle trading.* In this procedure it is important to observe that the total number of handles in the decomposition is *not* reduced but *stays constant.* However it allows one to replace an index 1 handle by an index 3 handle, dually an index $(n-1)$ handle by an index $(n-3)$ handle. So as long as $n \geq 5$ (which it is in the h-Cobordism Theorem) we will thus complete the verification of Proposition 2.42.

Here is how this trading procedure is done. Let h_1 be an index 1 handle in our decomposition. Since there is no 0-index handles, h_1 is attached directly to $\partial^- W \times [0,1]$. And we can hence use the fact that $\pi_1(W, \partial^- W) = 1$ to construct an embedded 2-disc $\mathbb{D}^2 \subseteq N = \partial^+ W_2$ such that $\mathscr{S} = \partial \mathbb{D}^2$ intersects the complementary sphere S to h_1 in exactly one point.

Next attach a canceling pair of handles h_2 and h_3 of indices 2 and 3, respectively, to $\partial^+ W_2$; i.e. so that

$$W_2 = W_2 + h_2 + h_3.$$

This can be done by the trivial inverse statement to Fact 3. And we can do it so that $\partial^- h_2 \cap \mathbb{D}^2 = \emptyset$ as well as $\partial^- h_2 \cap \partial^+ h = \emptyset$ for all the other handles h of index 2 in the given decomposition. At this point we have increased the number of handles in our decomposition by two; but we will shortly use h_2 to cancel h_1, thus trading h_1 for h_3.

To do this we need first to isotope the attaching sphere $\mathscr{S}_2$ of h_2 inside $\partial^+ W_2 = N$ so that its new attaching map $\hat{\mathscr{S}_2}$ is $\mathscr{S} \sharp \mathscr{S}_2$ as illustrated below in Figure 2.29.

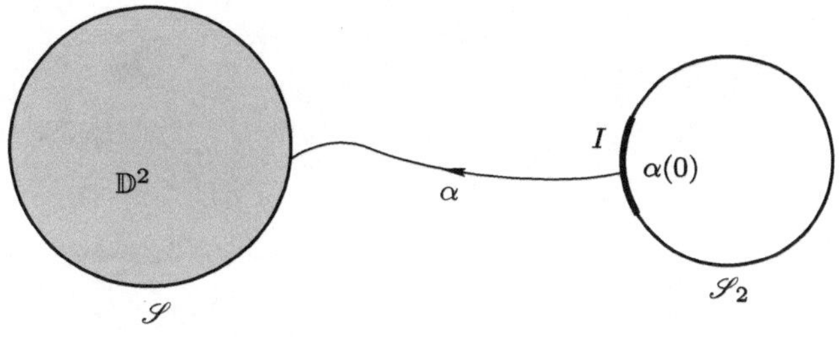

Figure 2.29

Here we have connected $\mathscr{S}_2$ to $\mathscr{S}$ by a smoothly embedded arc $\alpha\colon [0,1] \to N^{n-1}$ such that $\alpha(0) \in \mathscr{S}_2$, $\alpha(1) \in \mathscr{S}$ and $\alpha(t) \notin \mathbb{D}^2 \cup \mathscr{S}_2$ for all $t \in (0,1)$. And α is chosen so as to miss $\partial^+ h$ for all other index 2 handles h. The posited isotopy is now constructed by pushing a small closed interval I centered at $\alpha(0)$ (keeping its endpoints fixed) along α and across $\mathbb{D}^2$. After extending this to an isotopy of $\partial^- h_2$, we then use Fact 3 to cancel h_1 and h_2; thus giving a new decomposition with one fewer index 1 handle and one more index 3 handles. This finishes the verification of Proposition 2.42.

Remark 2.46. Freedman (1982) proved that every homotopy 4-sphere is homeomorphic to S^4. He did this by proving the following 5-dimensional version of the h-Cobordism Theorem.

Theorem 2.47. *Let W^5 be a smooth, simply-connected, 5-dimensional h-cobordism. Then W^5 is* homeomorphic *to $\partial^- W^5 \times [0,1]$.*

It is known that (in general) homeomorphism cannot be replaced by diffeomorphism in this theorem due to deep results of Donaldson. The key to this theorem is proving a topological version of Fact 4 when $n = 5$. This involves a complicated infinite procedure process invented by Casson and perfected by Freedman. It is this infinite process that causes the final result to be only a homeomorphism.

3

The s-Cobordism Theorem

3.1 Statement of the s-Cobordism Theorem

We now *drop* the first constraint (made in the h-Cobordism Theorem) that the h-cobordism W^n is *simply-connected*; but retain the second one that "$n \geq 6$". Then the arguments in §2.5 continue to work to show that W^n has a t-standard handlebody decomposition with no handles of indices 0, 1, $n-1$ and n. And an analogue of the trading handle technique works for higher index handles (in fact it is easier in this situation) to show that W^n has a t-standard handle decomposition with only handles of two consecutive indices i and $i+1$ where i is any chosen integer satisfying $2 \leq i \leq n-3$. For convenience of notation we will usually pick i to be 2; i.e. we can always assume that W^n has a t-standard handlebody decomposition with only handles of indices 2 and 3. Such a decomposition we will call a $(2,3)$-handlebody decomposition. In this case the associated chain complex C reduces to

$$D^3 \colon C_3 \to C_2.$$

Since the homology of C is identically zero (recall $H_s(C) = H_s(W, \partial^- W) = 0$) D^3 induces an isomorphism. Hence D^3 must be a square (say $m \times m$) matrix, and therefore

$$\text{the number of 2-handles} = \text{the number of 3-handles} = m.$$

We now fix some notation. Let $h_1, h_2, \cdots, h_m$ and $k_1, k_2, \cdots, k_m$ denote the index 2 and index 3 handles in our decomposition, respectively. Also let $\mathscr{S}_1, \mathscr{S}_2, \cdots, \mathscr{S}_m$ and $S_1, S_2, \cdots, S_m$ be the attaching and complementary spheres in N^{n-1} for the index 3 and index 2 handles, respectively. Where $N^{n-1} = \partial W_2$ is the codimension-1 submanifold of W^n which separates the index 2 from the index 3 handles. And using Fact 5 in §2.4, we can assume that

$$D^3 = I$$

is the identity $m \times m$ matrix with $\mathbb{Z}$-entries. This would seem to indicate that we can cancel the handles in pairs (h_i, k_i) using Fact 4' in §2.4; thus proving that $W = \partial^- W \times [0,1]$ without needing to assume that $\pi_1(W) = 1$. Unfortunately (or fortunately depending on one's point of view) this is not the case since the Whitney trick uses in an essential way that the loop $\gamma^{-1} * \alpha$ (§2.4, Fact 4) spans a (singular) 2-disc in N^{n-1}; i.e. is null homotopic. This is where the restriction $\pi_1(W) = 1$ was used in the proof of Smale's h-Cobordism Theorem. To proceed further without this assumption, we need some other reason for $\gamma^{-1} * \alpha$ to be null homotopic. For this we augment the information in the incidence matrix D^3 as follows. We define a new $m \times m$ matrix $\mathscr{D}$ with entries in a larger ring $\mathbb{Z}\Gamma$ where $\Gamma = \pi_1(W)$; which specializes to D^3 under the natural ring homomorphism $\mathbb{Z}\Gamma \to \mathbb{Z}$. Here $\mathbb{Z}\Gamma$ is the "integral group ring" of the group Γ which is defined as follows.

The elements of $\mathbb{Z}\Gamma$ are finite (formal) linear combination of elements in the group Γ with integral coefficients. Addition is the obvious one and multiplication is determined by the distributive law and the following formula for multiplying monomials:

$$(sg) \cdot (th) = (st)(gh)$$

where $s, t \in \mathbb{Z}$; $g, h \in \Gamma$ and $gh \in \Gamma$ is the product in the group Γ while $st \in \mathbb{Z}$ is the result of multiplication in the ring $\mathbb{Z}$.

$\mathbb{Z}\Gamma$ is an associative ring with identity, and is commutative if and only if Γ is an abelian group. It is in general a very complicated ring whose structure strongly reflects that of the group Γ. In fact it is hard to find examples of a pair of groups Γ and Π such that $\mathbb{Z}\Gamma \cong \mathbb{Z}\Pi$ but Γ is *not* isomorphic to Π. For pairs of finite groups Γ and Π, this was a well-known classical problem in finite group theory, due to G. Higman (in his Ph.D. thesis "Units in group rings", Oxford University, 1940. This problem was recently solved by M. Hertweck in [6], where he constructed the first such examples. In his example, $|\Gamma| = |\Pi| = 2^{21} \cdot 97^{28}$ and both are solvable groups.) There is a canonical ring epimorphism

$$\varepsilon \colon \mathbb{Z}\Gamma \to \mathbb{Z}$$

determined on monomials by $\varepsilon(sg) = s$, where $s \in \mathbb{Z}$ and $g \in \Gamma$.

To define $\mathscr{D}$, we fix base points $* \in N^{n-1}$, $\Delta_i \in S_i = S^{n-3}$ and $\Box_i \in \mathscr{S}_i = S^2$ for each index i, $1 \le i \le m$. We also choose arcs α_i and γ_i in N connecting $*$ to Δ_i and $\Box_i$, respectively. That is

$$\alpha_i(0) = \gamma_i(0) = *,$$

$$\alpha_i(1) = \Delta_i \quad \text{and} \quad \gamma_i(1) = \Box_i.$$

And for each (transverse) intersection point $x \in S_i \cap \mathscr{S}_j$ we connect x to Δ_i and $\Box_j$ by arcs

$$\delta_i^x \colon [0,1] \to S_i \quad \text{and} \quad \varepsilon_j^x \colon [0,1] \to \mathscr{S}_j$$

satisfying

$$\delta_i^x(0) = \varepsilon_j^x(0) = x,$$

$$\delta_i^x(1) = \Delta_i \quad \text{and} \quad \varepsilon_j^x(1) = \Box_j.$$

Then the concatenation

$$\gamma_j^{-1} * \varepsilon_j^x * (\delta_i^x)^{-1} * \alpha_i$$

is a loop in N^{-1} based at $*$ (Figure 3.1) and hence its homotopy class, denoted by $\beta_{ij}(x)$

$$\beta_{ij}(x) \colon = [\gamma_j^{-1} * \varepsilon_j^x * (\delta_i^x)^{-1} * \alpha_i] \in \pi_1(N) = \Gamma.$$

Then $\mathscr{D}_{ij}$ is defined by the following formula

$$\mathscr{D}_{ij} = \sum_{x \in S_i \cap \mathscr{S}_j} \sigma(x)\beta_{ij}(x) \in \mathbb{Z}\Gamma.$$

Recall $\sigma(x) = \pm 1$ is the orientation sign at the point of intersection x. (See §2.4.)

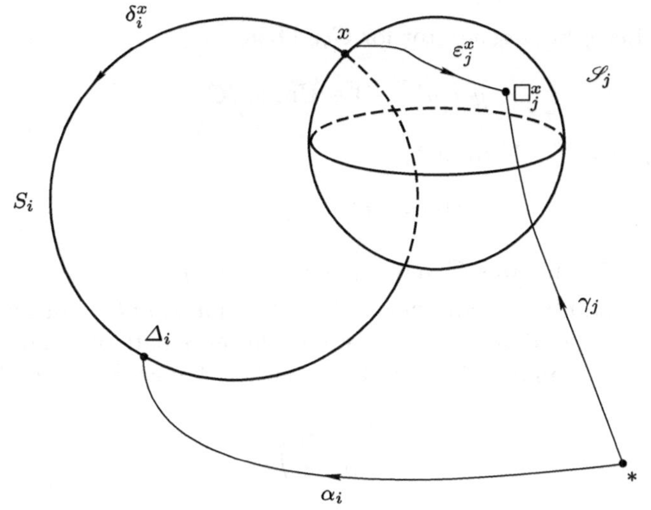

Figure 3.1

Clearly $\varepsilon(\mathscr{D}_{ij}) = D_{ij}^3$; therefore $\varepsilon(\mathscr{D}) = D^3$; i.e. $\mathscr{D}$ is an $m \times m$ matrix with entries in $\mathbb{Z}\Gamma$ which augments the $\mathbb{Z}$ data in D^3.

Let $p \colon \widetilde{W} \to W$ denote the universal cover of W; then $\widetilde{W}$ is a *proper h-cobordism* from $\partial^-\widetilde{W}$ to $\partial^+\widetilde{W}$ where $\partial^\pm\widetilde{W} = p^{-1}(\partial^\pm W)$. In particular

$$H_2(\widetilde{W}, \partial^-\widetilde{W}) = H_3(\widetilde{W}, \partial^-\widetilde{W}) = 0.$$

Elaborating on this it can be shown that $\mathscr{D}$ is an *invertible* $m \times m$-matrix over the ring $\mathbb{Z}\Gamma$; i.e. there exists a matrix $\mathscr{D}^{-1}$ with $\mathbb{Z}\Gamma$-entries such that

$$\mathscr{D}\mathscr{D}^{-1} = I = \mathscr{D}^{-1}\mathscr{D}.$$

Let us now examine to what extent the above choices of α_i, γ_j, etc. influence the matrix $\mathscr{D}$. Since S_i and $\mathscr{S}_j$ are simply-connected, they are spheres of dimension $n-3$ and 2 respectively, the choices of Δ_i, $\square_j$, δ_i^x and ε_j^x are irrelevant. But the choices of α_i and γ_j are relevant. Changing α_i changes the i-th row of $\mathscr{D}$ by multiplying it by an element of Γ, and by changing γ_j we change the j-th column of $\mathscr{D}$ by multiplication by an element of Γ. Also Fact 5 of §2.4 (slightly generalized) allows us to change $\mathscr{D}$ by adding any $\mathbb{Z}\Gamma$-multiple of some column of $\mathscr{D}$ to another (different) column. And of course we can permute columns or rows. And by changing the orientation of $\mathscr{S}_j$ or S_i we can multiply the j-th column or i-th row by -1. Summarizing we can change $\mathscr{D}$ by any finite sequence of elementary row or column operations with a single caveat; namely we can only multiply rows or columns by *special units* in the ring $\mathbb{Z}\Gamma$; i.e. we can only multiply by monomials $\pm\gamma$ where $\gamma \in \Gamma$.

Notation. Let C denote the infinite cyclic group and C_n denote the cyclic group of order n.

Example 3.1. Let g be a generator for C_5. Then

$$g + g^{-1} - 1 \in \text{Units}(\mathbb{Z}C_5);$$

its inverse is $g^2 + g^{-2} - 1$. In fact

$$\text{Units}(\mathbb{Z}C_5) = C \oplus C_{10}$$

where $g + g^{-1} - 1$ generates C and $-g$ generates C_{10}.

We can also modify $\mathscr{D}$ using "Fact 4" — the non-simply connected expansion of Fact 4 in §2.4 stated below — to either cancel a pair of handles or insert a trivial pair of $(2,3)$-handles. The effect is to either change $\mathscr{D}$ to the block diagonal $(m+1) \times (m+1)$-matrix

$$\begin{pmatrix} \mathscr{D} & 0 \\ 0 & 1 \end{pmatrix}$$

or vice-versa.

Fact 4. (non-simply connected Whitney trick) Consider $M^n + h_1 + h_2$ where $n \geq 6$, $I(h_1) = 2$ and $I(h_2) = 3$. If

$$\sum_{x \in \mathscr{S} \cap S} \sigma(x)\beta(x) = 1$$

then $M^n + h_1 + h_2 = M^n$.

We are about ready to state the generalization of Smale's h-Cobordism Theorem to non-simply connected manifolds due independently to Barden-Mazur-Stallings and called the s-Cobordism Theorem. For this we need the following preliminary definitions.

Definition 3.2. *An* elementary operation *on a square matrix is any of those described above (on last and this page). We say that a pair of invertible square matrices A and B (of perhaps different sizes) with entries in $\mathbb{Z}\Gamma$ are equivalent (written $A \equiv B$) if B can be gotten from A by a finite sequence of elementary operations. Let $[A]$ denote the equivalence class of A.*

Definition 3.3. *Two h-cobordisms W^n and W'^n with the same base (i.e. with $\partial^- W^n = \partial^- W'^n$) are called equivalent if there exists a diffeomorphism $f \colon W \to W'$ such that $f|_{\partial^- W} = \mathrm{id}$.*

s-Cobordism Theorem 3.4. *There is a bijective correspondence between the set of equivalence classes of smooth h-cobordisms with the same connected base M^{n-1} ($n \geq 6$) and the set of equivalence classes of invertible square matrices $\mathscr{D}$ with entries in $\mathbb{Z}\pi_1(M)$, denoted by $Wh(\pi_1 M)$; where the cylinder $M \times [0,1]$ corresponds to the class of the identity matrix I. In general an h-cobordism W with $\partial^- W = M^{n-1}$ corresponds to $[\mathscr{D}]$ where $\mathscr{D}$ is the intersection matrix for a $(2,3)$-handle decomposition of W^n.*

Remark 3.5. The integral group ring $\mathbb{Z}\Gamma$ is *universal* in the sense that given any ring R (with 1) and a group homomorphism

$$f \colon \Gamma \to \mathrm{Units}(R)$$

there is a *unique* extension

$$\overline{f} \colon \mathbb{Z}\Gamma \to R$$

of f to a ring homomorphism $\overline{f}$. Namely, $\overline{f}$ is determined by its values on monomials $sg \in \mathbb{Z}\Gamma$, where $s \in \mathbb{Z}$ and $g \in \Gamma$, by the formula

$$\overline{f}(sg) = \begin{cases} f(g) + f(g) + \cdots + f(g) & s \text{ summands}, \quad \text{if } s \geq 0, \\ (-f(g)) + \cdots + (-f(g)) & |s| \text{ summands}, \quad \text{if } s \leq 0. \end{cases}$$

To illustrate this observation, let us use it to show that the unit $u = t + t^{-1} - 1 \in \mathrm{Units}(\mathbb{Z}C_5)$, mentioned in Example 3.1, has infinite order. For this represent C_5 in $\mathrm{Units}(\mathbb{C})$ by f where $f(t) = e^{2\pi i/5} = \cos 72° + i \sin 72°$. Then

$$\overline{f}(u) = \overline{f}(t + t^{-1} - 1) = 2\cos 72° - 1 \in \mathbb{R}.$$

Using the inequalities

$$0 = \cos 90° < \cos 72° < \cos 60° = 1/2,$$

we see that $0 < |\overline{f}(u)| < 1$. Hence $\overline{f}(u)$ has infinite order in $\mathrm{Units}(\mathbb{C})$. Since $\overline{f}|_{\mathrm{Units}(\mathbb{Z}C_5)}$ is a group homomorphism of $\mathrm{Units}(\mathbb{Z}C_5)$ into $\mathrm{Units}(\mathbb{C})$, we conclude that u has infinite order in $\mathrm{Units}(\mathbb{Z}C_5)$.

Let us return now to the s-Cobordism Theorem. Recall that $Wh(\Gamma)$ denotes the set of all equivalence classes of invertible matrices with $\mathbb{Z}\Gamma$-coefficients.

The tools developed so far almost prove the s-Cobordism Theorem. The only major difficulty is showing that $[\mathscr{D}]$ is independent of the given $(2,3)$-handlebody decomposition of the h-cobordism (W^n, M^{n-1}). For this we need to discuss the *simple homotopy theory* of J. H. C. Whitehead. We will only do this briefly. (A more complete discussion is in Milnor's survey article [15].) However without developing this theory, we could prove the weaker statement that the set of equivalence classes of all h-cobordisms with the same base M^{n-1} ($n \geq 6$) is a quotient set of

$Wh(\pi_1 M)$; where the quotient set could depend on M^{n-1} (not just on $\pi_1 M^{n-1}$). Of course when $|Wh(\pi_1 M)| = 1$ there is no non-trivial quotient sets, and hence we would have already proven the s-Cobordism Theorem. This happens in the topologically very important special case when $\pi_1 M^{n-1}$ is a free abelian group. The case when $\pi_1 M^{n-1} = 1$ reduces, of course, to Smale's h-Cobordism Theorem.

3.2 Whitehead group

We start our discussion with Whitehead torsion by associating to each ring R (with 1) an abelian group $K_1(R)$ (in a functorial way). And we then show that $Wh(\Gamma)$ is in natural bijective correspondence with a certain quotient group of $K_1(\mathbb{Z}\Gamma)$. To R we can, of course, associate its group of units Units(R), or, more generally the group of units $GL_m(R)$ in the ring $M_m(R)$ of all $m \times m$ matrices with coefficients in R; i.e.

$$GL_m(R) = \text{Units}(M_m(R)).$$

As motivation recall that the incidence matrix $\mathscr{D}$ is an element of $GL_m(\mathbb{Z}\pi_1 M)$. And since $\mathscr{D}$ is equivalent to

$$\begin{pmatrix} \mathscr{D} & 0 \\ 0 & 1 \end{pmatrix}$$

we embed $GL_m(R)$ in $GL_{m+1}(R)$ by this group homomorphism which we call *stabilization*. But recall that we are only interested in the equivalence class of $\mathscr{D}$ which also allows certain elementary row and column operations to be performed on $\mathscr{D}$. We will show presently that this is the same (after stabilizing) as considering elements in a certain quotient group of the abelianization of $GL_m(\mathbb{Z}\pi_1 M)$. Given this background we define

$$K_1(R) = \lim_{m \to \infty} GL_m(R)/[GL_m(R), GL_m(R)],$$

or more concretely

$$K_1(R) = GL(R)/[GL(R), GL(R)]$$

where

$$GL(R) = \bigcup_{m=1}^{\infty} GL_m(R).$$

In this formula $GL_m(R)$ has been identified with a subgroup of $GL_{m+1}(R)$ by the above stabilization embedding.

Recall that $\pm\Gamma \subseteq \text{Units}(\mathbb{Z}\Gamma)$ denotes the subgroup consisting of the *special units* $\pm g$, where $g \in \Gamma$. Since $\text{Units}(\mathbb{Z}\Gamma) = GL_1(\mathbb{Z}\Gamma)$ we can identify $\pm\Gamma$ with a subgroup of $GL(\mathbb{Z}\Gamma)$ and we denote its image in $K_1(\mathbb{Z}\Gamma)$ by $\text{im}(\pm\Gamma)$.

Theorem 3.6. $Wh(\Gamma) = K_1(\mathbb{Z}\Gamma)/\,\text{im}(\pm\Gamma)$.

Remark 3.7. Hence the set $Wh(\Gamma)$ has a natural abelian group structure. In fact the assignment

$$\Gamma \to Wh(\Gamma)$$

is a functor from groups to abelian groups.

Remark 3.8. Let R be a commutative ring, then Units(R) is naturally a direct summand of $K_1(R)$. In fact

$$\det\colon GL_m(R) \to \mathrm{Units}(R)$$

splits the natural inclusion

$$\mathrm{Units}(R) = GL_1(R) \hookrightarrow GL_m(R)$$

and hence induces the desired splitting of Units(R) $\to K_1(R)$, which we also denote by

$$\det\colon K_1(R) \to \mathrm{Units}(R).$$

Let $SK_1(R) = \ker(\det)$. Then

$$K_1(R) = SK_1(R) \oplus \mathrm{Units}(R).$$

When Γ is an abelian group, $\mathbb{Z}\Gamma$ is a commutative ring and the above decomposition becomes

$$Wh(\Gamma) = \mathrm{Units}(\mathbb{Z}\Gamma)/(\pm\Gamma) \oplus SK_1(\mathbb{Z}\Gamma).$$

Example 3.9 (*Higman*). $SK_1(\mathbb{Z}C_5) = 0$ and hence

$$Wh(C_5) = \mathrm{Units}(\mathbb{Z}C_5)/(\pm C_5) = C \oplus (\pm C_5)/(\pm C_5) = C$$

where C is generated by $t + t^{-1} - 1$ with t being a generator of C_5.

More generally Bass (1964) showed that if p is a prime, then $SK_1(\mathbb{Z}C_p) = 0$ and hence

$$Wh(C_p) = \mathrm{Units}(\mathbb{Z}C_p)/(\pm C_p)$$

which is a free abelian group of rank $(p-3)/2$, or 0 if $p = 2$.

And Bass extended this to arbitrary finite groups G to yield that *$Wh(G)$ is a finitely generated abelian group of rank $r - q$ where*

$$r = \text{the number of conjugacy classes of pairs } \{g, g^{-1}\}, g \in G;$$
$$q = \text{the number of conjugacy classes of cyclic subgroups of } G.$$

Caveat 3.10. *In general, when G is finite, $Wh(G)$ can contain non-trivial elements of finite order, due to examples of R. Alperin, K. Dennis and M. Stein* (1973). *Also if Γ is not finite, $Wh(\Gamma)$ is frequently* not *even finitely generated, due to examples of Bass and Murthy* (1966).

Example 3.11. Let S_n be the symmetric group on n letters. One easily calculates the numbers r and q for S_n and sees that $r = q$. Hence by Bass' result $Wh(S_n)$ is a finite group. Later Oliver (1980) showed that, in fact, $Wh(S_n) = 0$.

Here is some preliminary notation needed to prove Theorem 3.6. Let $i \neq j$ be positive integers which are both smaller than or equal to some fixed integer m, and let $a \in R$. This data determines the $m \times m$ "elementary matrix" $e_{ij}(a)$ whose diagonal entries are all 1 and whose only nonzero off diagonal entry is a which occurs in the i-th row and j-th column of $e_{ij}(a)$. Note that $e_{ij}(a)$ is an invertible matrix whose inverse is $e_{ij}(-a)$.

Example 3.12.

$$e_{13}(2) = \begin{pmatrix} 1 & 0 & 2 \\ 0 & 1 & 0 \\ 0 & 0 & 1 \end{pmatrix}$$

where $m = 3$.

Hence $e_{ij}(a) \in GL_m(R) \subseteq GL(R)$, and we continue to use $e_{ij}(a)$ to denote its image in $GL(R)$. Let $E_m(R)$ denote the subgroup of $GL_m(R)$ generated by the elementary $m \times m$ matrices, and likewise $E(R)$ be the subgroup of $GL(R)$ generated by the images of all elementary matrices in $GL(R)$. The key to proving Theorem 3.6 is the following result due to Whitehead.

Lemma 3.13. $E(R) = [GL(R), GL(R)]$.

And this lemma is an immediate consequence of the following two results.

Proposition 3.14. *If $m \geq 3$, then $E_m(R) \subseteq [GL_m(R), GL_m(R)]$ and consequently $E(R) \subseteq [GL(R), GL(R)]$.*

Proof. Consider $e_{ij}(a)$ where $a \in R$ and $i, j \leq m$. Choose a positive integer $k \leq m$ which is different from both i and j. (This is where $m \geq 3$ is used.) Then

$$e_{ij}(a) = e_{ik}(a)e_{kj}(1)e_{ik}(-a)e_{kj}(-1) = [e_{ik}(a), e_{kj}(1)]. \qquad \square$$

Proposition 3.15. *$[GL_m(R), GL_m(R)] \subseteq E_{2m}(R)$ and consequently $[GL(R), GL(R)] \subseteq E(R)$.*

Proof. This is a direct consequence of the following three (easily verified) matrix identities in which $A, B \in GL_m(R)$:

1.
$$\begin{pmatrix} ABA^{-1}B^{-1} & 0 \\ 0 & I \end{pmatrix} = \begin{pmatrix} A & 0 \\ 0 & A^{-1} \end{pmatrix} \begin{pmatrix} B & 0 \\ 0 & B^{-1} \end{pmatrix} \begin{pmatrix} (BA)^{-1} & 0 \\ 0 & BA \end{pmatrix},$$

2.
$$\begin{pmatrix} A & 0 \\ 0 & A^{-1} \end{pmatrix} = \begin{pmatrix} I & A \\ 0 & I \end{pmatrix} \begin{pmatrix} I & 0 \\ I - A^{-1} & I \end{pmatrix} \begin{pmatrix} I & -I \\ 0 & I \end{pmatrix} \begin{pmatrix} I & 0 \\ I - A & I \end{pmatrix},$$

3.
$$\begin{pmatrix} I & A \\ 0 & I \end{pmatrix} = \prod_{i=1}^{m} \prod_{j=1}^{m} e_{i,j+m}(A_{ij}). \qquad \square$$

Remark 3.16. Multiplying $A \in GL_m(R)$ on the *left* by $e_{ij}(a)$, where $a \in R$, has the effect on A of the "elementary row operation" of multiplying the j-th row of A on the *left* by a and then adding the result to the i-th row of A. Likewise multiplying A on the right by $e_{ij}(a)$ effects A by the "elementary column operation" of multiplying the i-th column of A on the *right* by a and then adding the result to the j-th column of A.

Caveat 3.17. *Multiplying a row on the* right *and adding the result to a different row is* not *(in general) an elementary row operation.* Nor *is multiplying a column on the* left *and adding the result to a different column a column operation. To illustrate this consider the following exercise.*

Exercise 3.18. *Find a ring R (with 1) and elements $a, b, c, d, e \in R$ such that*

$$\begin{pmatrix} a & b \\ c & d \end{pmatrix} \in GL_2(R)$$

but

$$\begin{pmatrix} a + ce & b + de \\ c & d \end{pmatrix}$$

is not *invertible.*

Hint. Let $R = M_2(\mathbb{Z})$. Define $A, B \in \text{Units}(R)$ by

$$A = \begin{pmatrix} 1 & 1 \\ 0 & 1 \end{pmatrix}, \quad B = \begin{pmatrix} 1 & 0 \\ 1 & 1 \end{pmatrix}.$$

Let $a = I + AB$, $b = A$, $c = B$, $d = I$, $e = -A$. Then

$$\begin{pmatrix} a & b \\ c & d \end{pmatrix} = \begin{pmatrix} I & A \\ 0 & I \end{pmatrix} \begin{pmatrix} I & 0 \\ B & I \end{pmatrix} = e_{12}(A)e_{21}(B).$$

But

$$\begin{pmatrix} a + ce & b + de \\ c & d \end{pmatrix} = \begin{pmatrix} \begin{pmatrix} 2 & 0 \\ 0 & 0 \end{pmatrix} & 0 \\ B & I \end{pmatrix}.$$

Since $M_2(R) = M_4(\mathbb{Z})$

$$\begin{pmatrix} \begin{pmatrix} 2 & 0 \\ 0 & 0 \end{pmatrix} & 0 \\ B & I \end{pmatrix} = \begin{pmatrix} 2 & 0 & 0 & 0 \\ 0 & 0 & 0 & 0 \\ 1 & 0 & 1 & 0 \\ 1 & 1 & 0 & 1 \end{pmatrix}$$

is not invertible since its determinant is 0. □

3.3 Whitehead torsion for chain complexes

Unless stated otherwise a *ring* R means an *associative* ring with 1 but not necessarily *commutative.*

Recall that a (left) R-module is an abelian group M together with a multiplication $R \times M \to M$ of elements (vectors) in M by "scalars" in R satisfying the equations

$$\begin{aligned} a(x + y) &= ax + ay, \\ a(bx) &= (ab)x, \\ 1x &= x \end{aligned}$$

where $a, b \in R$ and $x, y \in M$.

An R-module M is *free* if it has a basis $\mathscr{B}$; i.e. a set of "vectors" such that any element $x \in M$ is "uniquely expressible" as a linear combination of elements in $\mathscr{B}$, with coefficients in R, where only finitely many coefficients are not 0. An R-module is *projective* if it is isomorphic to a direct summand of some free module. It is alternatively characterized as follows:

An R-module P is *projective* if and only if any diagram

$$\begin{array}{ccc} & & M \\ & & \downarrow f \\ P & \xrightarrow[g]{} & N \end{array}$$

where f and g denote R-module homomorphisms with f being surjective, then g factors through f; i.e. there is an R-module homomorphism $h\colon P \to M$ with $f \circ h = g$.

Our present task is to show how to associate to a "finite based" acyclic chain complex C consisting of free R-modules

$$C:\ C_n \overset{d_n}{\to} C_{n-1} \overset{d_{n-1}}{\to} \cdots \to C_1 \overset{d_1}{\to} C_0$$

an element $\tau(C) \in \overline{K}_1(R)$ where $\overline{K}_1(R)$ is the quotient group $K_1(R)/\pm 1$, called the "reduced group" of $K_1(R)$. (Here ± 1 denotes the subgroup of $K_1(R)$ which is the image of the "trivial units" $\pm 1 \in \mathrm{Units}(R)$.) By "finite based" we mean that each chain module C_i is equipped with a *finite* basis β_i and that $n < \infty$. In doing this we make the following extra (very weak) ongoing assumption about the ring R.

Standard Assumption. If β and β' are a pair of finite bases for the same R-module M having cardinality m and n, respectively, then $m = n$.

Remark 3.19. For any group Γ its integral group ring $\mathbb{Z}\Gamma$ satisfies the standard assumption; as does any ring R that admits a homomorphism to a commutative ring. In the case of $\mathbb{Z}\Gamma$, this homomorphism is the augmentation homomorphism $\varepsilon\colon \mathbb{Z}\Gamma \to \mathbb{Z}$.

The importance of introducing $\overline{K}_1(R)$ is that the element in this group determined by a square matrix A is unchanged if we permute either the rows or columns of A. This is a direct consequence of the following reduction of the transposition

$$\begin{pmatrix} 0 & 1 \\ 1 & 0 \end{pmatrix}$$

to

$$\begin{pmatrix} -1 & 0 \\ 0 & 1 \end{pmatrix}$$

by the indicated sequence of elementary row and column operations

$$\begin{pmatrix} 0 & 1 \\ 1 & 0 \end{pmatrix} \to \begin{pmatrix} 0 & 1 \\ 1 & 1 \end{pmatrix} \to \begin{pmatrix} -1 & 1 \\ 0 & 1 \end{pmatrix} \to \begin{pmatrix} -1 & 0 \\ 0 & 1 \end{pmatrix}.$$

Now let $b = \{b_1, b_2, \cdots, b_m\}$ and $c = \{c_1, c_2, \cdots, c_m\}$ be a pair of bases for a free R-module M. (Note that we have used here our standard assumption to conclude that b and c have the same number of members m.) Let A be the matrix determined by the change of basis from b to c; i.e. by the set of m linear combinations

$$c_i = \sum_{j=1}^{m} A_{ij} b_j.$$

We denote the corresponding element in $\overline{K}_1(R)$ determined by A with $[c/b]$. Note that $[c/b]$ is independent of the ordering of b and c used to define A. This is why we used $\overline{K}_1(R)$ instead of $K_1(R)$. Note that if d is a third basis of M, then a straightforward calculation shows the important formula

$$[d/b] = [d/c] + [c/b]. \tag{3.1}$$

We will use Eq. (3.1) to show that the following formula gives a well-defined homomorphism

$$\tau\colon Aut(M) \to \overline{K}_1(R).$$

Let $f \in Aut(M)$ be an R-module automorphism of the free R-module M and b be a finite basis for M, then

$$\tau(f) = [f(b)/b].$$

To show that $\tau(f)$ is well defined, let c be a second basis for M. Then we need to show that

$$[f(b)/b] = [f(c)/c].$$

By applying Eq. (3.1) twice to the bases $f(c)$, $f(b)$, b and c, we obtain the equation

$$[f(c)/c] = [f(c)/f(b)] + [f(b)/b] + [b/c].$$

So it suffices to show that

$$[f(c)/f(b)] = -[b/c].$$

For this first note that $[c/b] = -[b/c]$ which is seen by applying Eq. (3.1) to the three bases c, b, c; i.e.

$$1 = [c/c] = [c/b] + [b/c].$$

Therefore it suffices to show that

$$[f(c)/f(b)] = [c/b].$$

But this is a consequence of the fact that the matrix A not only determines $[c/b]$ but also determines $[f(c)/f(b)]$ as is seen from the following easy calculation:

$$f(c_i) = f\left(\sum_{j=1}^{m} A_{ij} b_j\right) = \sum_{j=1}^{m} A_{ij} f(b_j).$$

To show that τ is a group homomorphism, let g be a second R-module automorphism of M and apply Eq. (3.1) to the bases $g(f(b))$, $f(b)$ and b which yields

$$\tau(gf) = [g \circ f(b)/b] = [g(f(b))/f(b)] + [f(b)/b] = \tau(g) + \tau(f).$$

Definition 3.20. *$\tau(f)$ is called the* torsion *of the automorphism f.*

Remark 3.21. With more effort we can factor τ through $K_1(R)$. Fix an *ordered* basis $b = (b_1, b_2, \cdots, b_m)$ for M. Then $f(b) = (f(b_1), f(b_2), \cdots, f(b_m))$ is a second *ordered* basis and the change of basis matrix $A(f)$ defined by the equations

$$f(b_i) = \sum_{j=1}^{m} A(f)_{ij} b_j$$

gives a group isomorphism $f \mapsto A(f)$ between $Aut(M)$ and $GL_m(R)$. Composing this isomorphism with the natural homomorphism $GL_m(R) \to K_1(R)$ induces the factorization

$$\widetilde{\tau}\colon Aut(M) \to K_1(R).$$

Remark 3.22. The proof that $[f(c)/f(b)] = [c/b]$ is easily extended to verify this identity where $f\colon M \to N$ is any isomorphism of R-modules from the case where $f \in Aut(M)$; i.e. $M = N$.

We conclude the discussion by showing how to extend the definition of $\tau\colon Aut(M) \to \overline{K}_1(R)$ from finitely generated *free* R-modules M to finitely generated *projective* R-modules P. We start by noticing that P is a direct summand of a finitely generated free R-module. This is seen by picking a set of generators $a_1, a_2, \cdots, a_n$ for P and defining an R-module surjection $p\colon R^n \twoheadrightarrow P$ by $p(b_i) = a_i$ where $\{b_1, b_2, \cdots, b_n\}$ is a basis for R^n. Since p *splits*, P is isomorphic to a direct summand of R^n.

Therefore we can find finitely generated R-modules Q and M such that M is free and

$$P \oplus Q = M.$$

Embed the group $Aut(P)$ into $Aut(M)$ by sending $f \in Aut(P)$ to $f \oplus \mathrm{id}_Q \in Aut(M)$ and compose this embedding with the group homomorphism $\tau\colon Aut(M) \to \overline{K}_1(R)$. This is our definition of

$$\tau\colon Aut(P) \to \overline{K}_1(R).$$

It is a pleasant exercise to show that τ is independent of the choices of Q and M made above.

Hint. Use $\widetilde{K}_0(R)$ and Remark 3.22. □

We say that a pair of bases b, c for V are equivalent (denoted $b \sim c$) iff $[b/c] = 0$. Using Eq. (3.1) one easily sees that if $b \sim b'$ and $c \sim c'$ are bases for V, then $[c/b] = [c'/b']$.

If $U = V \oplus W$ where b and c are bases for V and W, respectively, then we let bc denote the basis for U obtained by stringing together these two bases. More generally, if

$$0 \to V \to U \xrightarrow{p} W \to 0$$

is a short exact sequence of free R-modules, and c' is a *lifting* of c to U, then the concatenation bc' is a basis for U. By a lifting of c, we mean a subset c' of U such that $p\colon c' \to c$ is a bijection of sets. Now one easily shows that the equivalence class bc' depends only on the equivalence classes of b and c, respectively.

Now we are ready to define the torsion $\tau(C) \in \overline{K}_1(R)$ for a finitely based acyclic chain complex of free R-modules

$$C\colon\ C_n \overset{d_n}{\to} C_{n-1} \to \cdots \to C_1 \overset{d_1}{\to} C_0.$$

For this we temporarily make the following, a bit annoying, assumption.

Assumption 3.23. *Each* $B_i = \mathrm{im}(d_{i+1})$ *is a free* R*-module.*

Let c_i be the given basis for C_i. Then

$$\tau(C) = \sum_{i=0}^{n} (-1)^n [c_i/b_i] \tag{3.2}$$

where b_i is a second basis for C_i determined as follows. Choose a basis β_i for B_i arbitrarily, and consider the short exact sequence

$$0 \to B_i \to C_i \to B_{i-1} \to 0$$

defined since $\ker(d_i) = \mathrm{im}(d_{i+1}) = B_i$. Then $b_i = \beta_i \beta'_{i-1}$. It is a pleasant exercise to show that $\tau(C)$ is well-defined.

In the important case where C has length two; i.e. $C_i = 0$ for $i \neq n$ and $n-1$, Assumption 3.23 *always* holds and Eq. (3.2) specializes to become

$$\tau(C) = (-1)^{n-1} [d_n(c_n)/c_{n-1}] \tag{3.3}$$

which is seen as follows. First note that $d\colon C_n \to C_{n-1}$ is an isomorphism; hence $B_n = 0$ and $B_{n-1} = C_{n-1}$. So we can choose $\beta_{n-1} = d_n(c_n)$ and $\beta'_{n-1} = c_n$. With these choices b_n becomes c_n and $b_{n-1} = d_n(c_n)$. And Eq. (3.2) becomes

$$\tau(C) = (-1)^n [c_n/c_n] + (-1)^{n-1} [d_n(c_n)/c_{n-1}] = (-1)^{n-1} [d_n(c_n)/c_{n-1}]$$

verifying Eq. (3.3).

To get around our Assumption 3.23 in general, one can proceed as Milnor does in [15, §4] via introducing the notion of "stable bases". But instead we indicate an alternative approach which has since then also become popular.

We start by recalling that a free chain complex is acyclic if and only if it is "contractible"; i.e. there exists a chain contraction (i.e. chain homotopy of id to 0)

$$\sigma = \{\sigma_i\colon C_i \to C_{i+1}\}$$

satisfying $d_{i+1} \circ \sigma_i + \sigma_{i-1} \circ d_i = \mathrm{id}_{C_i}$ for all i. Fix such a chain contraction and define

$$C_{\mathrm{even}} = \bigoplus_{i=0}^{\infty} C_{2i} \quad \text{and} \quad C_{\mathrm{odd}} = \bigoplus_{i=0}^{\infty} C_{2i+1}.$$

(Of course these are only finite sums.) Then define $\Sigma_{\text{even}}\colon C_{\text{even}} \to C_{\text{odd}}$ by

$$\Sigma_{\text{even}} = \bigoplus_{i=0}^{\infty} \sigma_{2i}$$

and $D_{\text{even}}\colon C_{\text{even}} \to C_{\text{odd}}$ by

$$D_{\text{even}} = \bigoplus_{i=0}^{\infty} d_{2i}.$$

Similarly we define $\Sigma_{\text{odd}}\colon C_{\text{odd}} \to C_{\text{even}}$ and $D_{\text{odd}}\colon C_{\text{odd}} \to C_{\text{even}}$.

Lemma 3.24. *$f = D_{\text{even}} + \Sigma_{\text{even}}\colon C_{\text{even}} \to C_{\text{odd}}$ is invertible.*

Proof. Consider the composition $f \circ g$ where

$$g = D_{\text{odd}} + \Sigma_{\text{odd}}\colon C_{\text{odd}} \to C_{\text{even}}.$$

We will show that $f \circ g\colon C_{\text{odd}} \to C_{\text{odd}}$ is invertible and consequently so is f. Now

$$\begin{aligned} f \circ g &= D_{\text{even}} \circ D_{\text{odd}} + \Sigma_{\text{even}} \circ \Sigma_{\text{odd}} + (D_{\text{even}} \circ \Sigma_{\text{odd}} + \Sigma_{\text{even}} \circ D_{\text{odd}}) \\ &= 0 + \Sigma_{\text{even}} \circ \Sigma_{\text{odd}} + \text{id}. \end{aligned}$$

And note that $\Sigma_{\text{even}} \circ \Sigma_{\text{odd}}$ is nilpotent since it increases degree by 2 and C has finite dimension; i.e. $f \circ g = I - N$ where $N^m = 0$ for some $m \in \mathbb{Z}^+$. But

$$(I - N)(I + N + N^2 + \cdots + N^{m-1}) = I - N^m = I.$$

Therefore $f \circ g$ is invertible. □

And since both C_{even} and C_{odd} are finitely based free R-modules, $\tau(f) \in \overline{K}_1(R)$ is defined as above where C_{even} and C_{odd} have dimension 1 and 0, respectively. (Recall Assumption 3.23 holds for all length 2 complexes.) So our general definition is

$$\tau(C) = [f(c_{\text{even}})/c_{\text{odd}}] \tag{3.4}$$

where $c_{\text{even}} = c_0 c_2 \cdots$, $c_{\text{odd}} = c_1 c_3 \cdots$ and $f = D_{\text{even}} + \Sigma_{\text{even}}$.

We now proceed to show that this new general definition (3.4) of $\tau(C)$ agrees with our old one using Eq. (3.2) when it is applicable; i.e. when Assumption 3.23 holds. For this we explicitly construct the chain contraction σ. Recall that the basis b_i for C_i used in Eq. (3.2) is $b_i = \beta_i \beta'_{i-1}$ where β_i is a basis for $B_i = \text{im}(d_{i+1})$ and that $d_i\colon \beta'_{i-1} \to \beta_{i-1}$ is a set bijection. Let $\delta_i\colon \beta_{i-1} \to \beta'_{i-1}$ denote its inverse and let B'_{i-1} be the free submodule of C_i spanned by β'_{i-1}. Therefore $C_i = B_i \oplus B'_{i-1}$ and we define

$$\sigma_i\colon C_i = B_i \oplus B'_{i-1} \to C_{i+1} = B_{i+1} \oplus B'_i$$

by specifying it on the basis b_i for C_i as follows:

$$\sigma_i|_{\beta'_{i-1}} \equiv 0 \quad \text{and} \quad \sigma_i|_{\beta_i} = \delta_{i+1}.$$

Clearly σ_i maps B_i isomorphically to B'_i and is identically zero on B'_{i-1}. Its matrix representation in terms of the bases b_i for C_i and b_{i+1} for C_{i+1} has the block form shown in Figure 3.2, while $d_{i+1}\colon C_{i+1} \to C_i$ has the block matrix form shown in Figure 3.3.

$$\begin{array}{c|c|c|} & \beta_i & \beta'_{i-1} \\ \hline \beta_{i+1} & 0 & 0 \\ \hline \beta'_i & I & 0 \end{array}$$

Figure 3.2

$$\begin{array}{c|c|c|} & \beta_{i+1} & \beta'_i \\ \hline \beta_i & 0 & I \\ \hline \beta'_{i-1} & 0 & 0 \end{array}$$

Figure 3.3

Using this one easily sees that $\sigma_{i-1}d_i + d_{i+1}\sigma_i = \mathrm{id}_{C_i}$; i.e. that σ is a chain contraction of C.

And also the following equations result immediately from these matrix description of d_{i+1} and σ_i:

$$d_i(\beta_i) = 0, \quad d_i(\beta'_{i-1}) = \beta_{i-1}, \quad \sigma_i(\beta_i) = \beta'_i, \quad \sigma_i(\beta'_{i-1}) = 0.$$

After recalling that $f = D_{\text{even}} + \Sigma_{\text{even}}$ we deduce from these that

$$\begin{aligned} f(b_{2i}) &= f(\beta_{2i}\beta'_{2i-1}) \\ &= f(\beta_{2i})f(\beta'_{2i-1}) \\ &= \sigma_{2i}(\beta_{2i})d_{2i}(\beta'_{2i-1}) \\ &= \beta'_{2i}\beta_{2i-1}. \end{aligned}$$

And using this we easily see that

$$f(b_{\text{even}}) = b_{\text{odd}} \tag{3.5}$$

where $b_{\text{even}} = b_0b_2b_4\cdots$ and $b_{\text{odd}} = b_1b_3b_5\cdots$. Now consider the following calculation

$$\begin{aligned} [f(c_{\text{even}})/c_{\text{odd}}] &= [f(c_{\text{even}})/f(b_{\text{even}})] + [f(b_{\text{even}})/b_{\text{odd}}] + [b_{\text{odd}}/c_{\text{odd}}] \\ &= [c_{\text{even}}/b_{\text{even}}] + [b_{\text{odd}}/b_{\text{odd}}] - [c_{\text{odd}}/b_{\text{odd}}] \end{aligned}$$

in which Eq. (3.5) was used to identify the middle term as $[b_{\text{odd}}/b_{\text{odd}}]$. Hence

$$[f(c_{\text{even}})/c_{\text{odd}}] = \sum_{i=1}^{\infty}(-1)^i[c_i/b_i].$$

Since the left side term in this identity is the definition of $\tau(C)$ using Eq. (3.4) and the right side term is the definition using Eq. (3.2), this verifies that the two definitions agree when Eq. (3.2) is applicable.

But it remains to show that definition (3.4) is a "good definition"; i.e. independent of the particular chain contraction σ used. For this let $\hat{\sigma}$ be a second chain contraction and $\hat{f} = D_{\text{even}} + \hat{\Sigma}_{\text{even}}$ where $\hat{\Sigma}_{\text{even}} = \oplus_{i=0}^{\infty}\hat{\sigma}_{2i}$. We need to show that

$$[\hat{f}(c_{\text{even}})/c_{\text{odd}}] = [f(c_{\text{even}})/c_{\text{odd}}].$$

And since

$$0 = [c_{\text{odd}}/c_{\text{odd}}] = [c_{\text{odd}}/\hat{f}(c_{\text{even}})] + [\hat{f}(c_{\text{even}})/f(c_{\text{even}})] + [f(c_{\text{even}})/c_{\text{odd}}]$$

this is equivalent to show that

$$[f^{-1} \circ \hat{f}(c_{\text{even}})/c_{\text{even}}] = [\hat{f}(c_{\text{even}})/f(c_{\text{even}})] = 0. \tag{3.6}$$

To show this we will need the following two facts.

Fact 1. If A is a strictly upper triangular $m \times m$ matrix (i.e. its diagonal entries are all 1) then the class of A, denoted $[A]$, in $\overline{K}_1(R)$ is 0.

Fact 2. Let $V = P \oplus Q$ be a finitely generated free R-module and $\eta\colon V \to V$ be an R-module endomorphism such that $\eta(Q) \subseteq P$ and $\eta(P) = 0$. Then $\varphi = \text{id}_V + \eta$ is an automorphism of V (i.e. $\text{id}_V + \eta \in Aut(V)$) and $\tau(\text{id}_V + \eta) = 0$ in $\overline{K}_1(R)$.

To verify Fact 1 just observe that A can be changed to I by a finite sequence of elementary row operations.

To verify Fact 2 first observe that $\varphi = \text{id}_V + \eta$ is an automorphism of V since $\text{id}_V - \eta$ is its inverse. To see that $\tau(\text{id}_V + \eta) = 0$ in $\overline{K}_1(R)$, first stabilize φ to $\hat{\varphi} = \varphi \oplus \text{id} \in Aut(V \oplus V)$. Then

$$V \oplus V = (P \oplus Q) \oplus (P \oplus Q)$$

and pick a basis $b = b_1, b_2, \cdots, b_m, b_{m+1}, \cdots b_{2m}$ for $V \oplus V$ such that

$$b_1, b_2, \cdots, b_m \in P \oplus 0 \oplus 0 \oplus Q, \quad b_{m+1}, b_{m+2}, \cdots, b_{2m} \in 0 \oplus Q \oplus P \oplus 0.$$

And note that the matrix A which represents $\hat{\varphi}$ with respect to this basis is strictly upper triangular. Hence $[A] = 0$ in $\overline{K}_1(R)$ by Fact 1. But

$$[A] = [\hat{\varphi}(b)/b] = \tau(\hat{\varphi}) = \tau(\varphi).$$

Hence $\tau(\varphi) = 0$.

To use these facts in verifying that Eq. (3.6) is true, we need a nice formula for f^{-1}. We saw before that f^{-1} is "almost" $D_{\text{odd}} + \Sigma_{\text{odd}}$. But if σ is "carefully chosen" as in the verification of Eqs. (3.2) and (3.4), then $\sigma_i \circ \sigma_{i-1} = 0$ and hence

$$f^{-1} = D_{\text{odd}} + \Sigma_{\text{odd}}.$$

(For this we observe that to define σ_i in this way, it is only necessary that all B_j are projective R-modules; being free is not needed.)

This choice having been made

$$f^{-1} \circ \hat{f} = D_{\text{odd}} D_{\text{even}} + (\Sigma_{\text{odd}} D_{\text{even}} + D_{\text{odd}} \hat{\Sigma}_{\text{even}}) + \Sigma_{\text{odd}} \hat{\Sigma}_{\text{even}}. \qquad (3.7)$$

Note that $D_{\text{odd}} D_{\text{even}} = 0$ and let α and ω denote $\Sigma_{\text{odd}} \hat{\Sigma}_{\text{even}}$ and $\hat{\Sigma}_{\text{even}} - \Sigma_{\text{even}}$, respectively. With these notations Eq. (3.7) becomes

$$f^{-1} \circ \hat{f} = I + D_{\text{odd}} \omega + \alpha.$$

Next observe that

$$D_{\text{odd}} \omega = \omega' D_{\text{odd}} \qquad (3.8)$$

where $\omega' = \hat{\Sigma}_{\text{odd}} - \Sigma_{\text{odd}}$. This is a consequence of the fact that both σ and $\hat{\sigma}$ are chain contraction of d.

We now apply Fact 2 with $V = C_{\text{even}}$ and $\eta = D_{\text{odd}} \omega$, $P = \ker(D_{\text{even}})$ and $Q = \operatorname{im}(\Sigma_{\text{odd}})$. Using Eq. (3.8), we see that $\eta(Q) \subset P$ and $\eta(P) = 0$. Hence $I - \eta = (I + \eta)^{-1}$ and $\tau(I + \eta) = 0$ in $\overline{K}_1(R)$.

Note that verifying Eq. (3.6) is equivalent to showing that $\tau(f^{-1} \circ \hat{f}) = 0$ where $f^{-1} \circ \hat{f} \in Aut(V)$. Recall that $\tau \colon Aut(V) \to \overline{K}_1(R)$ is a group homomorphism and observe that

$$\begin{aligned} f^{-1} \circ \hat{f} &= (1 + \eta)(1 - \eta)(1 + \eta + \alpha) \\ &= (1 + \eta)(1 + (1 - \eta)\alpha). \end{aligned}$$

Hence

$$\tau(f^{-1} \circ \hat{f}) = \tau(1 + \eta) + \tau(1 + (1 - \eta)\alpha) = \tau(1 + (1 - \eta)\alpha).$$

Now note that $(1 - \eta)\alpha \colon C_{2i} \to C_{2i+2}$; hence the matrix A representing $1 + (1 - \eta)\alpha$ with respect to the "ordered basis" c_{even} is *strictly lower triangular.* And consequently

$$\tau(1 + (1 - \eta)\alpha) = 0$$

by the obvious variant of Fact 1.

The proof of Fact 2, combined with the extension of τ from free to projective modules, can be easily extended to yield the following

Fact 2'. Let $0 \to P_1 \to P_2 \to P_3 \to 0$ be a short exact sequence of finitely generated projective R-modules and let $f_i \in Aut(P_i)$ be such that the two sequences in the following ladder commute

$$\begin{array}{ccccc} P_1 & \longrightarrow & P_2 & \longrightarrow & P_3 \\ \downarrow{\scriptstyle f_1} & & \downarrow{\scriptstyle f_2} & & \downarrow{\scriptstyle f_3} \\ P_1 & \longrightarrow & P_2 & \longrightarrow & P_3 \end{array}$$

then $\tau(f_2) = \tau(f_1) + \tau(f_3)$ in $\overline{K}_1(R)$.

And the following important result can be deduced from Fact 2'.

Corollary 3.25. *Let $0 \to C' \to C \to C'' \to 0$ be a short exact sequence of finitely based free acyclic R chain complexes. Assume that the bases are consistent; i.e. $c_i \sim c_i' c_i''$ for each i, where c_i, c_i' and c_i'' are the given bases of C_i, C_i' and C_i'', respectively. Then*

$$\tau(C) = \tau(C') + \tau(C'').$$

We conclude this section by some remarks concerning Assumption 3.23.

We will show that B_i is always *stably free*; i.e. $B_i \oplus F$ is free for some finitely generated free R-module F. To see this argue as follows. We first show, by induction on i, that each B_i is a projective R-module. This is obvious for $i = 0$ since $B_0 = C_0$. Assume now that B_i is projective and consider the following diagram

$$\begin{array}{ccc} & & C_{i+1} \\ & & \big\downarrow{\scriptstyle d_{i+1}} \\ B_i & \xrightarrow{\text{id}} & B_i \end{array}$$

Since B_i is projective, we can lift id to $\eta_i \colon B_i \to C_{i+1}$; i.e. *split* d_{i+1}. Let $B_i' = \text{im}(\eta_i)$ and recall that $\ker(d_{i+1}) = \text{im}(d_{i+2}) = B_{i+1}$; consequently

$$B_{i+1} \oplus B_i' = C_{i+1}. \tag{3.9}$$

Therefore B_{i+1} is projective which verifies the inductive step.

Observe also that each B_i is finitely generated since it is the image of C_{i+1}. Therefore B_i determines a class

$$[B_i] \in \widetilde{K}_0(R) \stackrel{\text{def}}{=} K_0(R)/\text{classes of free modules}.$$

And note that a finitely generated projective R-module P is *stably free* if and only if $[P] = 0$ in $\widetilde{K}_0(R)$. Now observe that Eq. (3.9) yields

$$[B_{i+1}] + [B_i] = [C_{i+1}] = 0$$

since B_i' is isomorphic to B_i and C_{i+1} is free. Therefore

$$[B_{i+1}] = -[B_i] \tag{3.10}$$

and we see, by induction on i, that $[B_{i+1}] = 0$; where Eq. (3.10) gives the inductive step, and the induction starts with $[B_0] = [C_0] = 0$.

4

Some Classical Results

4.1 Novikov's Theorem

Let $G_{n,n+k}$ be the space of all n-dimensional subspaces of $\mathbb{R}^{n+k}$. Two subspaces U, V are close if the "angle between them" is small; i.e. let $\hat{U} = U \cap S^{n+k-1}$ and $\hat{V} = V \cap S^{n+k-1}$, then U is ε-close to V if and only if $\hat{V} \subseteq \varepsilon$-neighborhood of $\hat{U}$ in S^{n+k-1} (and also $\hat{U} \subseteq \varepsilon$-neighborhood of $\hat{V}$).

Remark 4.1. It can be shown that $G_{n,n+k}$ is a closed smooth nk-dimensional manifold.

The natural embedding of $\mathbb{R}^{n+k}$ into $\mathbb{R}^{n+k+1}$ induces an identification of $G_{n,n+k}$ with a subspace of $G_{n,n+k+1}$ and we define

$$G_n = \bigcup_{k=1}^{\infty} G_{n,n+k}$$

which can be thought of as all n-dimensional subspaces of $\mathbb{R}^\infty = \bigcup_{m=1}^{\infty} \mathbb{R}^m$.

Now an n-dimensional vector bundle E^n with base a compact Hausdorff space X defines a map $f_E \colon X \to G_n$ as follows. By Swan's identification of vector bundles with projective $C(X)$-modules there exists a k-dimensional vector bundle $\mathscr{E}^k \to X$ such that

$$E^n \oplus \mathscr{E}^k = \Theta^{n+k} = X \times \mathbb{R}^{n+k} \xrightarrow{q} \mathbb{R}^{n+k}.$$

For each $x \in X$, let E_x be the fiber in E^n over x. Then $q(E_x) \subset \mathbb{R}^{n+k}$ is an n-dimensional subspace of $\mathbb{R}^{n+k}$ and hence of $\mathbb{R}^\infty$. This subspace is $f_E(x)$. Let $[f_E] \in [X, G_n]$ denote its homotopy class.

Proposition 4.2. *This correspondence defines a bijection between isomorphism classes of n-dimensional vector bundles with base X and homotopy classes of continuous maps $[X, G_n]$.*

Remark 4.3. In the literature G_n is also dented by $BO(n)$ because of the proposition above. Here $O(n)$ denotes the group of orthogonal motions of $\mathbb{R}^n$.

With more work we can define a "classifying space" $BTop(n)$ for fiber bundles with fiber $\mathbb{R}^n$ in the sense of the following proposition.

Proposition 4.4. *There is a natural bijective correspondence between isomorphism classes of $\mathbb{R}^n$-bundles with base X and homotopy classes of continuous maps $[X, BTop(n)]$. Furthermore there is a natural map*

$$BO(n) \xrightarrow{\varphi} BTop(n)$$

that corresponds to forgetting the extra structure in an n-dimensional vector bundle and recalling only that it is an $\mathbb{R}^n$-bundle; i.e. if E^n is an n-dimensional vector bundle and $f_E \colon X \to BO(n)$ classifies it as a vector bundle, then $\varphi \circ f_E \colon X \to BTop(n)$ classifies it as an $\mathbb{R}^n$-bundle.

Novikov's Theorem 4.5. *Let $E^n \to S^m$ be an n-dimensional vector bundle $(n > m)$ which is trivial as an $\mathbb{R}^n$-bundle. Then it is trivial as a vector bundle; i.e.*

$$\pi_m BO(n) \xrightarrow{\varphi_*} \pi_m BTop(n)$$

is injective if $m < n$.

Note that an n-dimensional vector bundle E^n can be "stabilized" by adding the trivial line bundle Θ^1 to it; i.e.

$$E^n \to E^n \oplus \Theta^1 \to (E^n \oplus \Theta^1) \oplus \Theta^1 \to \cdots .$$

This allows us to identify $BO(n)$ with a subspace of $BO(n+1)$. Let

$$BO = \bigcup_{n=1}^{\infty} BO(n).$$

Likewise we can form

$$BTop = \bigcup_{n=1}^{\infty} BTop(n)$$

and φ induces a natural map $\Phi \colon BO \to BTop$.

Now $[X, BO]$ can be identified with $\widetilde{KO^0}(X)$ where the abelian group structure on $\widetilde{KO^0}(X)$ corresponds to an H-space structure $BO \times BO \to BO$ on BO, which corresponds to fiberwise direct sum $\oplus$ of vector bundles (with the same base). Similarly $BTop$ has an H-space structure and Φ respects these two H-spaces structures. Also $[X, BTop] = \widetilde{K}^0_{\mathrm{Top}}(X)$ which is the result of the Grothendieck construction performed to the stable isomorphism classes of $\mathbb{R}^n$-bundles over X.

Now Novikov's Theorem can be reinterpreted to say that

Theorem 4.6. *$\Phi_* \colon \pi_i(BO) \to \pi_i(BTop)$ is monic for each i.*

Using results in algebraic topology developed in 1950's (perhaps most importantly Serre's mod $\mathscr{C}$ extension of the Hurewicz and Whitehead Theorems (1953)) it can be deduced from Theorem 4.6 and the fact that Φ is an H-spaces map, that

Corollary 4.7. *$\Phi^* \colon H^i(BTop; \mathbb{Q}) \to H^i(BO; \mathbb{Q})$ is an epimorphism.*

Corollary 4.7 is the form of Theorem 4.6 which is usually referred to as Novikov's Theorem; i.e. his 1970 Field's medal result. Let us examine this in a bit more detail.

A (stable) rational characteristic class for vector bundles is by definition an element of $H^*(BO;\mathbb{Q})$. Now it is known that $H^*(BO;\mathbb{Q})$ is a (graded) polynomial ring on generators $p_1, p_2, \cdots$ with $p_i \in H^{4i}(BO;\mathbb{Q})$ which is called the i-th universal Pontryagin class. (See [17] for details.) If E is a vector bundle over X, then its i-th Pontryagin class $p_i(E) \in H^{4i}(X;\mathbb{Q})$ is $f_E^*(p_i)$ where $f_E\colon X \to BO$ classifies E (stably). And the i-th (rational) Pontryagin class of a smooth manifold M, denoted $p_i(M)$, is by definition $p_i(TM)$. It is clear that if $g\colon M \to N$ is a diffeomorphism, then $p_i(M) = g^*p_i(N)$, since $g^*(TN) = TM$ in $\widetilde{KO^0}(M)$ under this assumption. However if g is *only a homeomorphism*, this is no longer clear. However Kister-Mazur-Milnor had shown that $g^*(TN) = TM$ in $\widetilde{K}^0_{\mathrm{Top}}(M)$; i.e. the following diagram commutes (up to homotopy)

$$\begin{array}{ccccc} M & \xrightarrow{f_{TM}} & BO & \xrightarrow{\Phi} & BTop \\ \downarrow{\scriptstyle g} & & & \nearrow{\scriptstyle \Phi} & \\ N & \xrightarrow{f_{TN}} & BO & & \end{array}$$

Using Corollary 4.7, let $\hat{p}_i \in H^{4i}(BTop;\mathbb{Q})$ map to p_i under Φ^*. Therefore

$$f_{TM}^*\Phi^*(\hat{p}_i) = g^*f_{TN}^*\Phi^*(\hat{p}_i),$$

hence $f_{TM}^*(p_i) = g^*f_{TN}^*(p_i)$ and $p_i(M) = g^*p_i(N)$. This is the most familiar version of Novikov's result; i.e.

Theorem 4.8 (Novikov). *The rational Pontryagin classes are homeomorphism invariants.*

Although g^*TN needs not equal TM, as it does when g is a diffeomorphism, Novikov's Theorem implies that $TM - g^*TN$ represents an element of finite order in $KO^0(M)$ when g is a homeomorphism and M is compact. This follows immediately from the fact that if E_1 and E_2 are two n-dimensional vector bundles with the same compact Hausdorff base space X and $p_i(E_1) = p_i(E_2)$ in $H^{4i}(X;\mathbb{Q})$ for all i, then $[E_1] - [E_2] \in KO^0(X)$ has finite order, where $[E_1]$ and $[E_2]$ are the classes in $KO^0(X)$ represented by these bundles.

4.2 A counterexample to the Hurewicz Conjecture

We will now use Novikov's Theorem (Theorem 4.8) to construct a counterexample to the Hurewicz Conjecture 1.5. We start by recalling the Bott Periodicity Theorem. (Actually a much more elementary result is all that is needed, see Remark 4.16). Bott Periodicity Theorem is the following calculation of $\widetilde{KO^0}(S^n)$.

Theorem 4.9 (Bott).

$$\widetilde{KO^0}(S^n) = \begin{cases} \mathbb{Z}, & \textit{if } n \equiv 0 \pmod 4, \\ \mathbb{Z}_2, & \textit{if } n \equiv 1, 2 \pmod 8, \\ 0, & \textit{otherwise.} \end{cases}$$

In particular $\widetilde{KO^0}(S^4) = \mathbb{Z}$. Hence we have

Lemma 4.10. *There exists a 6-dimensional vector bundle $\mathbb{R}^6 \to E \to S^4$ such that $p_1(E) \neq 0$ in $H^4(S^4; \mathbb{Q})$.*

Remark 4.11. Note that every element $x \in \widetilde{KO^0}(B)$ is represented by a vector bundle. A priori $x = E_1 - E_2$ (formal difference), but recall that there exists an E_3 such that $E_2 \oplus E_3 = \Theta^m$ whose class is 0 in $\widetilde{KO^0}(B)$. Therefore $E_3 = -E_2$ and x is represented by $E_1 \oplus E_3$.

To verify Lemma 4.10, by this remark we can represent any non-zero multiple sx of a generator x of $\widetilde{KO^0}(S^4) = \mathbb{Z}$ by a vector bundle $\mathbb{R}^6 \to E_s \to S^4$. We claim that $p_1(E_s) \neq 0$. If not, then $p_i(E_s) = 0$ for all i since $H^{4i}(S^4; \mathbb{Q}) = 0$ for $i > 1$. And consequently sx would have finite order, which is a contradiction.

Now it is also known (See Steenrod's book [20]) that m-dimensional vector bundles with base S^n are in natural bijective correspondence with $\pi_{n-1}O(m)$ via the "clutching construction"; i.e. if $x \mapsto A_x$ ($x \in S^{n-1}$, $A_x \in O(m)$) is a representative, let

$$E = \mathbb{D}^n \times \mathbb{R}^m \sqcup_A \mathbb{D}^n \times \mathbb{R}^m = \mathbb{D}^n \times \mathbb{R}^m \sqcup \mathbb{D}^n \times \mathbb{R}^m / \equiv$$

where $(x, v) \equiv (x, A_x(v))$ for $x \in S^{n-1}$, $v \in \mathbb{R}^m$. And let $S(E) = \mathbb{D}^n \times S^{m-1} \sqcup_A \mathbb{D}^n \times S^{m-1}$ be the associated sphere bundle $S^{m-1} \to S(E) \to S^n$. Note that $S(E)$ is a closed smooth $(n + m - 1)$-dimensional manifold. We proceed to show that

Lemma 4.12. *We can find a positive integer s such that $S(E_s)$ is homotopy equivalent to $S^4 \times S^5$.*

Assuming this momentarily, then we show that $M = S(E_s)$ and $N = S^4 \times S^5$ are a pair of closed simply-connected smooth 9-dimensional manifolds which are homotopy equivalent but *not* homeomorphic, thus establishing one of Novikov's counterexamples to the Hurewicz Conjecture.

Note that $N = \partial(S^4 \times \mathbb{D}^6)$ and that $S^4 \times \mathbb{D}^6 = (S^4 \times [-1, 1]) \times \mathbb{D}^5 \subset \mathbb{R}^5 \times \mathbb{R}^5$ is a codimension-0 submanifold of $\mathbb{R}^{10}$ and hence parallelizable. Therefore $T(S^4 \times S^5) \oplus \Theta^1 = \Theta^9$ and hence TN represents 0 in $\widetilde{KO^0}(N)$. An elementary argument shows for any vector bundle $E \xrightarrow{q} B$ whose base is a smooth manifold that $q^*E \oplus q^*TB$ is equivalent to TE in $\widetilde{KO^0}(E)$. And clearly $TS(E) \oplus \Theta^1 = i^*(TE)$, where $i \colon S(E) \to E$ is the inclusion. Let $\hat{q} = q \circ i \colon S(E) \to B$ be the composite, consequently $\hat{q}^*(TB) \oplus \hat{q}^*E = TS(E)$ in $\widetilde{KO^0}(S(E))$. Apply these considerations to $E_s \xrightarrow{q} S^4$ and notice that $TS^4 \oplus \Theta^1 = \Theta^5$, we see that $\hat{q}^*(E_s) = TM$ in

$\widetilde{KO}^0(M)$. But $\hat{q}$ has a cross section $\sigma\colon S^4 \to E_s$, since $\dim E_s > \dim S^4$. Therefore $\sigma^*(TM) = \sigma^*\hat{q}^*(E_s) = E_s$, which has infinite order in $\widetilde{KO}^0(S^4)$. Consequently TM has infinite order in $\widetilde{KO}^0(M)$.

But if $g\colon M \to N$ is a homeomorphism, then by Novikov's Theorem 4.8 $TM - g^*TN = TM$ would have have finite order. This contradiction establishes that M and N are *not* homeomorphic.

It remains to verify Lemma 4.12; i.e. to show that for certain $s \in \mathbb{Z}_+$, $S(E_s)$ is homotopy equivalent to $S^4 \times S^5$. For this we must also, instead of $O(m)$, consider the larger H-space $G(S^{m-1})$ consisting of all degree ± 1 self maps of S^{m-1}. The clutching construction can also be done using $\pi_{n-1}G(S^{m-1})$ instead of $\pi_{n-1}O(m)$. The result is a spherical fibration over S^n with fiber S^{m-1}. And the natural map $\pi_{n-1}O(m) \to \pi_{n-1}G(S^{m-1})$ (when $m > n+1$) induced by the inclusion $O(m) \subset G(S^{m-1})$ is called the stable J-homomorphism (discovered by G. W. Whitehead). If $J(\alpha) = 0$, where $\alpha \in \pi_{n-1}O(m)$, then $S(E_\alpha)$ is homotopy equivalent to $S^n \times S^{m-1} = N$. But G. W. Whitehead showed that

$$\pi_{n-1}G(S^{m-1}) = \pi_{n+m-1}(S^m)$$

when $m > n+1$ and Serre proved the following

Serre's Theorem 4.13.

$$\pi_i(S^k) = \begin{cases} \mathbb{Z}, & \text{if } i = k, \\ \mathbb{Z} \oplus \text{finite group}, & \text{if } i = 2k-1 \text{ and } k \text{ is even}, \\ \text{finite group}, & \text{otherwise}. \end{cases}$$

Remark 4.14. Serre received the Field's medal in 1954 for this result.

Now apply the above analysis to the bundle E_s. Let $\alpha_s \in \pi_3 O(6) \cong \mathbb{Z}$ be the element corresponding to $s \in \mathbb{Z}_+$. Look at the J-homomorphism

$$J\colon \pi_3 O(6) \to \pi_9 S^6,$$

the target group is finite by Serre's Theorem (isomorphic to $\mathbb{Z}_{24}$, in fact). Therefore there exists an $s > 0$ such that $J(\alpha_s) = 0$. For such s, $S(E_s) \simeq S^4 \times S^5$.

We finish this section with some remarks on the topology of $SO(4)$, and a partial analysis of $\pi_3(O(5))$ which is independent of Bott's Theorem 4.9.

Remark 4.15. $SO(4)$ is diffeomorphic to $SO(3) \times S^3$ and $SO(3)$ is diffeomorphic to $\mathbb{R}\mathrm{P}^3$.

To see that $SO(3) \cong \mathbb{R}\mathrm{P}^3$, define $q\colon S^3 \subset \mathbb{H} \to SO(3)$ by $q(a)(x) = axa^{-1}$, where x is a "pure imaginary" unit length quaternion; i.e. $x \in S = \{r\mathbf{i}+s\mathbf{j}+t\mathbf{k} \mid r^2+s^2+t^2 = 1\}$ and note that $|a||b| = |ab|$ for $a, b \in \mathbb{H}$. Hence the transformation $b \mapsto aba^{-1}$ is in $SO(4)$. But 1 is an eigenvector for this transformation, hence it takes S to itself. q is a Lie group homomorphism whose kernel is $\{\pm 1\}$ since the center of $\mathbb{H}$ is $\mathbb{R}$. Since both S^3 and $SO(3)$ are closed 3-manifolds and q is an open map, it is onto, and $SO(3) = S^3/\pm 1 = \mathbb{R}\mathrm{P}^3$.

There is a principal fiber bundle $SO(3) \hookrightarrow SO(4) \xrightarrow{p} S^3$, where $p\colon SO(4) \to S^3$ is given by sending $A \in SO(4)$ to $A \cdot 1 \in S^3$. We can also define a smooth map $l\colon S^3 \to SO(4)$, sending $a \in S^3$ to the orthogonal transformation l_a of $\mathbb{R}^4 = \mathbb{H}$ induced by the left multiplication by a. Then it's easy to see $l\colon S^3 \to SO(4)$ is a cross section of the principal fiber bundle $SO(3) \hookrightarrow SO(4) \xrightarrow{p} S^3$, which is hence trivial. Therefore $SO(4) \cong SO(3) \times S^3$.

Remark 4.16. Consider the fiber bundle $O(4) \to O(5) \to S^4$. It induces a long exact sequence of homotopy groups

$$\cdots \to \pi_4(O(5)) \to \pi_4(S^4) \to \pi_3(O(4)) \to \pi_3(O(5)) \to \pi_3(S^4) \to \cdots .$$

Notice that $\pi_3(S^4) = 0$, $\pi_4(S^4) = \mathbb{Z}$, $\pi_3 O(5) = \pi_4(BO) = \widetilde{KO^0}(S^4)$, which is the group of stable vector bundles over S^4 (the group structure comes in 2 ways, from π_4 and from the H-space structure on BO, hence coincide). Also since $SO(4) \cong S^3 \times \mathbb{R}\mathrm{P}^3$ (homeomorphic as spaces), we have $\pi_3(O(4)) = \mathbb{Z} \oplus \mathbb{Z}$. $\pi_4(O(5))$ is the set of isomorphism classes of 5-dimensional vector bundles over S^5, and the map $\pi_4(O(5)) \to \pi_4(S^4)$ is given by taking the Euler number $\chi(E)$ of such a vector bundle E. Let σ be the 0-section of E, make σ transverse to itself and count the intersections with orientation, this is just $\chi(E^5)$. For odd dimension n and n-dimensional vector bundles $E^n \to S^n$, $\chi(E^n)$ is always 0. Hence we see that $\pi_3 O(5) = \mathbb{Z} \oplus$finite group. And by considering the homotopy exact sequence of the fiber bundle $O(5) \to O(6) \to S^5$ one easily sees that $\pi_3(O(6)) = \pi_3(O(5))$. This partial calculation is enough to make the above construction of the closed manifold $S(E_s) \simeq S^4 \times S^5$ but not homeomorphic to $S^4 \times S^5$. Namely, let x in the previous argument be any generator for a $\mathbb{Z}$ summand of $\pi_3(O(6))$ and the argument goes through as before.

4.3 Milnor's exotic spheres

Now there are two 4-dimensional vector bundles over S^4 that arise naturally: TS^4, the tangent bundle and η, the canonical quaternionic line bundle over the quaternionic projective line $\mathbb{HP}^1$, which is the space of all quaternionic lines in $\mathbb{H}^2$ (see [17, p.244])

$$\eta = \{v \in L |\ L \in \mathbb{HP}^1\}.$$

We are interested in finding the Euler class e and the first Pontryagin class p_1 for these bundles.

Since η has a complex structure, it is naturally oriented. Let $u \in H^4(T(\eta))$ be the Thom class of η. Since $S^4 = \mathbb{HP}^1 \subset \mathbb{HP}^2 = T(\eta)$ — the Thom space of η, we can consider u to be an element in $H^4(S^4)$. Now since $e(\eta) \cup u = u^2$ in $H^8(T(\eta))$, we see that $e(\eta) = u$. Now orient S^4 so that $\langle u, [S^4]\rangle = 1$. This also orients TS^4. Since $\langle e(TS^4), [S^4]\rangle = \chi(S^4) = 2$, we see that $e(TS^4) = 2u$.

Next we determine p_1. It is also clear that $p_1(TS^4) = 0$ since p_1 is a stable characteristic class and $TS^4 \oplus \Theta^1 = \Theta^5$. However, the determination of $p_1(\eta)$ is more sophisticated. It involves the definition of p_1 in terms of Chern classes; i.e. $p_1(\eta) = -c_2(\eta \otimes \mathbb{C})$. By the product formula for Chern classes,

$$c_2(\eta \otimes \mathbb{C}) = c_2(\eta \oplus \overline{\eta}) = c_2(\eta) + c_1(\eta)c_1(\overline{\eta}) + c_2(\overline{\eta}) = 2c_2(\eta) = 2e(\eta),$$

where there is a canonical complex structure on η, and $\overline{\eta}$ is its complex conjugate. Therefore we have $p_1(\eta) = -2u$.

Let $\hat{T}, \hat{\eta} \in \pi_3 SO(4)$ be the classes corresponding to TS^4 and η, respectively.

Caveat 4.17. *$\hat{T} + \hat{\eta} \neq T \oplus \eta$, since the addition in $\pi_3 SO(4) = \pi_3(G_4)$ is* not *directly related to the addition in $KO^0(S^4)$, i.e. Whitney sum of vector bundles.*

But the following is true.

Lemma 4.18. *Let E_α be the 4-dimensional vector bundle over S^4, corresponding to $\alpha \in \pi_3 SO(4)$, then the following maps*

$$p_1 \colon \pi_3 SO(4) \to \mathbb{Z}, \quad \alpha \mapsto \langle p_1(E_\alpha), [S^4] \rangle,$$

$$\chi \colon \pi_3 SO(4) \to \mathbb{Z}, \quad \alpha \mapsto \langle e(E_\alpha), [S^4] \rangle$$

are homomorphisms of abelian groups. Especially we have

$$\langle p_1(a\hat{T} + b\hat{\eta}), [S^4] \rangle = a\langle p_1(\hat{T}), [S^4] \rangle + b\langle p_1(\hat{\eta}), [S^4] \rangle = -2b$$

and

$$\chi(a\hat{T} + b\hat{\eta}) = a\chi(\hat{T}) + b\chi(\hat{\eta}) = 2a - b$$

for any $a, b \in \mathbb{Z}$.

We will prove this lemma after using it to prove the main result.

Theorem 4.19. *There exists a 4-dimensional vector bundle E over S^4 such that*

1. *$S(E)$ is homeomorphic to S^7;*
2. *$S(E)$ is* not *diffeomorphic to S^7;*
3. *$\mathbb{D}(E)/S(E)$ is a closed topological manifold which does* not *support a smooth structure.*

We start our proof with the following result.

Lemma 4.20. *If a 4-dimensional vector bundle E over S^4 has $\chi(E) = 1$, then $S(E)$ is homeomorphic to S^7.*

Proof. By Smale's proof of the GPC it suffices to show that $S(E)$ is homotopy equivalent to S^7. Using the long exact sequence of homotopy groups for the fiber bundle $S^3 \to S(E) \xrightarrow{q} S^4$, we see that $\pi_1 S(E) = 0$ as is $\pi_2 S(E)$. Hence $H_1(S(E)) = H_2(S(E)) = 0$ as are $H_5(S(E))$ and $H_6(S(E))$ by Poincaré duality. Also we know that $H_0(S(E)) = H_7(S(E)) = \mathbb{Z}$. Now the Gysin-Wang sequence yields

$$0 \to H^3(S(E)) \to H^0(S^4) \xrightarrow{\cup \chi(E) u} H^4(S^4) \xrightarrow{q^*} H^4(S(E)) \to 0,$$

where u is a generator of $H^4(S^4)$. Clearly $\chi(E) = 1$ yields $H^3(S(E)) = H^4(S(E)) = 0$ and by Poincaré duality $H_3(S(E)) = H_4(S(E)) = 0$. And by Hurewicz Theorem $\pi_7 S(E) = \mathbb{Z}$. Let $f \colon S^7 \to S(E)$ represent a generator, then since f_* is an isomorphism on homology, it is a homotopy equivalence by Whitehead Theorem. □

For the remainder of our proof of Theorem 4.19, we will assume that $\chi(E) = 1$. Hence (1) is satisfied, and clearly (3) will imply (2).

To prove (3), with certain extra constraints on $p_1(E)$, we consider the following 8-dimensional case of the Hirzebruch Index Theorem (due to Thom).

Index Formula 4.21. *Let M^8 be a smooth oriented closed 8-dimensional manifold. Then*

$$I(M^8) = \langle \frac{1}{45}(7p_2(M) - p_1(M)^2), [M^8]\rangle.$$

We will assume that $M^8 = \mathbb{D}(E)/S(E)$ is smooth and show this constrains $p_1(E)$. Note that M^8 is the Thom complex of E and by the Thom Isomorphism Theorem (or by Lefschetz duality) $\widetilde{H}^{i+4}(M^8) = H^i(S^4)$. Hence

$$H^i(M^8) = \begin{cases} \mathbb{Z}, & \text{if } i = 0, 4, 8, \\ 0, & \text{otherwise} \end{cases}$$

and $I(M^8) = \pm 1$. By changing orientation we may assume that $I(M^8) = 1$. Then the index formula becomes

$$45 = 7\langle p_2(M), [M]\rangle - \langle p_1(M)^2, [M]\rangle,$$

therefore

$$\langle p_1(M)^2, [M]\rangle \equiv 4 \pmod 7. \tag{4.1}$$

Now $E \subset M$ is an open submanifold and hence a smooth submanifold. (Recall that $E = M - \infty$.)

Caveat 4.22. *But the induced smooth structure α on E from M may be different from the one β coming as a vector bundle over S^4. However here is where Novikov's Theorem comes to rescue; i.e. $p_1(E) \in H^4(E; \mathbb{Q})$ is the same element regardless of the smooth structure. And we know how to calculate the one coming from β.*

Recall $TE = q^*(E \oplus TS^4) = q^*(E) \oplus q^*(TS^4)$, therefore $TE \oplus \Theta^1 = q^*(E) \oplus \Theta^5$, hence $p_1(TE) = q^*(p_1(E))$. Let $\sigma\colon E \subset M$ be the inclusion, and note that $\sigma^*\colon H^4(M; \mathbb{Z}) \to H^4(E; \mathbb{Z})$ is an isomorphism and $\sigma^*(p_1(M)) = p_1(TE)$. Therefore $\sigma^*(p_1(M)) = q^*(p_1(E))$. Let $\tau\colon S^4 \to E$ be the zero section, then $\tau^*\sigma^*(p_1(M)) = p_1(E)$ and we have $\langle p_1(M), \sigma_*\tau_*[S^4]\rangle = \langle p_1(E), [S^4]\rangle \in \mathbb{Q}$. Thus $p_1(M) = \langle p_1(E), [S^4]\rangle \cdot u$, where u is a generator of $H^4(M; \mathbb{Z})$ such that $\langle u, [S^4]\rangle = 1$.

So the congruence relation (4.1) becomes

$$\langle \langle p_1(E), [S^4]\rangle^2 u^2, [M]\rangle \equiv 4 \pmod 7.$$

But $\langle u^2, [M]\rangle = 1$ (since we have chosen the orientation of M such that $I(M) = 1$), therefore $\langle p_1(E), [S^4]\rangle^2 \equiv 4 \pmod 7$. This implies

$$\langle p_1(E), [S^4]\rangle \equiv 2 \text{ or } 5 \pmod 7. \tag{4.2}$$

We proceed to use Lemma 4.18 to find a bundle E such that $\chi(E) = 1$ and $\langle p_1(E), [S^4]\rangle \neq 2$ or $5 \pmod 7$; i.e. $\langle p_1(E), [S^4]\rangle \equiv 0, 1, 3, 4, 6 \pmod 7$. We can achieve any of these congruences; but let us focus on achieving 0.

Since $E \mapsto \hat{E}$ is a bijection, we let $E_{a,b}$ be such that $\hat{E}_{a,b} = a\hat{T} + b\hat{\eta} \in \pi_3 SO(4)$. Then $\chi(E_{a,b}) = 2a - b$ and $\langle p_1(E_{a,b}), [S^4]\rangle = -2b$. Since we want $\chi(E_{a,b}) = 1$, we require $2a - b = 1$, therefore $\langle p_1(E_{a,b}), [S^4]\rangle = -2(2a-1) = 2 - 4a$, which we want to be congruent to 0 modulo 7; i.e. $7|4a - 2$. If we let $a = 4$ and hence $b = 7$, this works. (Actually the Chinese remainder Theorem shows any congruence modulo 7 can be achieved.) Now $E = E_{4,7}$ satisfies statement (3) of Theorem 4.19; i.e. $\mathbb{D}(E_{4,7})/S(E_{4,7})$ is a closed topological 8-manifold without a smooth structure.

Finally we give the proof of Lemma 4.18.

Proof of Lemma 4.18. Let G_4^+ be the space of oriented 4-planes in $\mathbb{R}^\infty$, which is a model for $BSO(4)$. Then there is a canonical isomorphism $\pi_3 SO(4) \cong \pi_4 G_4^+$, under which $\alpha \in \pi_3 SO(4)$ corresponds to the classifying map $f_\alpha\colon S^4 \to G_4^+$ of E_α. Under this identification,

$$p_1\colon \pi_3 SO(4) \to \mathbb{Z}$$

is given by

$$p_1(\alpha) = \langle p_1(E_\alpha), [S^4]\rangle = \langle p_1, f_{\alpha*}[S^4]\rangle = \langle p_1, h(f_\alpha)\rangle,$$

which is clearly a homomorphism, where $h\colon \pi_4(G^+) \to H_4(G^+)$ is the Hurewicz homomorphism and $p_1 \in H^4(G_4^+)$ is the universal first Pontryagin class.

The same argument applies to $\chi\colon \pi_3 SO(4) \to \mathbb{Z}$, using the universal Euler class of an oriented 4-dimensional vector bundle $e \in H^4(G_4^+)$. (This works even though e is *not* a stable class.) □

4.4 Rochlin's Theorem

Rochlin's Theorem 4.23. *Let M^4 be a smooth closed almost parallelizable 4-manifold, then the index of M, $I(M)$ is divisible by* 16.

Proof. Note first that the almost parallelizability implies that M is orientable, choose an orientation for M, then $I(M)$ is defined. Recall the index formula

$$I(M) = \frac{1}{3}\langle p_1(M), [M]\rangle$$

therefore $3I(M) = \langle p_1(M), [M]\rangle \in \mathbb{Z}$. Hence the theorem is equivalent to showing that $\langle p_1(M), [M]\rangle$ is divisible by 48.

By definition, M is almost parallelizable means that $M - \mathrm{int}\mathbb{D}^4$ is parallelizable, as is $\mathbb{D}^4$. Let $\sigma_1, \cdots, \sigma_4$ and $e_1, \cdots, e_4$ be (orthonormal) framings for $T(M - \mathrm{int}\mathbb{D}^4)$ and $T\mathbb{D}^4$ respectively, where $e_1, \cdots, e_4$ is the standard framing for $\mathbb{D}^4 \subset \mathbb{R}^4$; i.e. $e_1 = (1, 0, 0, 0)$, etc. Then

$$TM = \mathbb{D}^4 \times \mathbb{R}^4 \bigcup_{S^3 \times \mathbb{R}^4} (M - \mathrm{int}\mathbb{D}^4) \times \mathbb{R}^4$$

where $S^3 \times \mathbb{R}^4 \subset \mathbb{D}^4 \times \mathbb{R}^4$ is glued to $S^3 \times \mathbb{R}^4 \subset (M - \text{int}\mathbb{D}^4) \times \mathbb{R}^4$ by $(x, v) \mapsto (x, A_x(v))$ with $A_x \in SO(4)$. In fact A_x is the matrix representing the basis $\sigma_1(x), \cdots, \sigma_4(x)$ in terms of $e_1, \cdots, e_4$. Note that $A\colon S^3 \to SO(4)$. Hence we can form a 4-plane bundle ξ over S^4 by the "clutching construction" using A; i.e.

$$\xi = \mathbb{D}^4_+ \times \mathbb{R}^4 \bigcup_{S^3 \times \mathbb{R}^4} \mathbb{D}^4_- \times \mathbb{R}^4$$

where $(x, v) = (x, A_x(v))$ for $x \in S^3$ and $v \in \mathbb{R}^4$. Let $S^3 \times [0,1]$ be a collar neighborhood of $S^3 \subset \partial(M - \text{int}\mathbb{D}^4)$ in $W = M - \text{int}\mathbb{D}^4$. Define $f\colon M \to S^4$ by $f\colon \mathbb{D}^4 \to \mathbb{D}^4_+$, $f(S^3 \times [0,1]) = \mathbb{D}^4_-$ and $f(W - S^3 \times [0,1]) =$ the south pole of $\mathbb{D}^4_-$ (Figure 4.1).

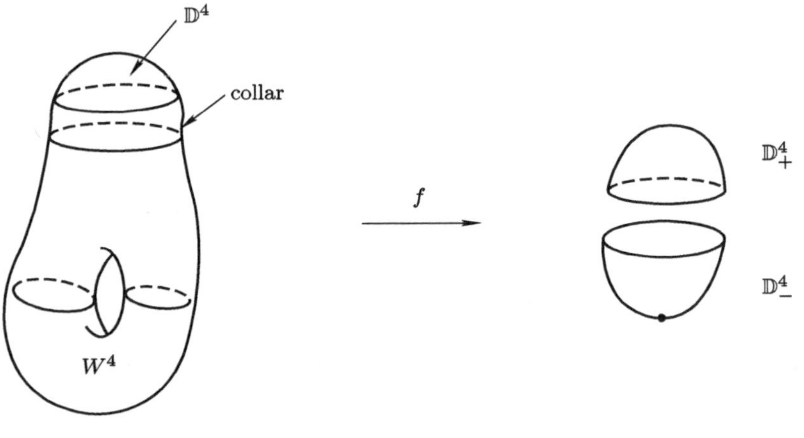

Figure 4.1

There is clearly a bundle map $\hat{f}\colon TM^4 \to \xi$ covering $f\colon M \to S^4$; i.e. $\hat{f}$ is a linear isomorphism between the fiber of TM^4 over each point $x \in M^4$ to the fiber of ξ over the corresponding point $f(x) \in S^4$, i.e. $TM^4 = f^*(\xi)$. Hence $f^*(p_1(\xi)) = p_1(TM) = p_1(M)$. Therefore

$$\begin{aligned} \langle p_1(M^4), [M^4]\rangle &= \langle f^*(p_1(\xi)), [M^4]\rangle \\ &= \langle p_1(\xi), f_*[M^4]\rangle \\ &= \langle p_1(\xi), [S^4]\rangle \end{aligned}$$

since f is clearly a degree-one map. So Rochlin's Theorem 4.23 is equivalent to showing that $\langle p_1(\xi), [S^4]\rangle$ is divisible by 48.

The next thing to observe is that ξ is *not* an arbitrary bundle over S^4, but that when considered as a stable bundle $J(\xi \oplus \Theta^1) = 0$. Recall $J\colon \pi_i SO(n) \to \pi_{n+i}(S^n)$ and that $\xi \in \pi_3(SO(4))$. But stably $\xi \oplus \Theta^1 \in \pi_3 SO(5)$, so $J\colon \pi_3 SO(5) \to \pi_8(S^5)$, where $\pi_3 SO(5) \cong \mathbb{Z}$ (by Bott) and $\pi_8(S^5) \cong \mathbb{Z}_{24}$ (by Serre).

Assuming momentarily the fact that $J(\xi \oplus \Theta^1) = 0$, we complete the proof of Rochlin's Theorem. Recall from last section that the map

$$\pi_3 SO(5) = \pi_4(G_5) \to \mathbb{Z},\ E \mapsto \langle p_1(E), [S^4] \rangle$$

is a group homomorphism. Let E_0 be a generator of the infinite cyclic group $\pi_3 SO(5)$. Then $\xi \oplus \Theta = sE_0$. Hence $\langle p_1(\xi), [S^4] \rangle = s\langle p_1(E_0), [S^4] \rangle$. Now it is known (see the reference in Remark 4.25, or more generally due to Adams' J-homomorphism papers) that $J\colon \pi_3 SO(5) \to \pi_8(S^5)$ is surjective. Thus $s = 24t$ for $t \in \mathbb{Z}$, because $J(\xi \oplus \Theta^1) = 0$. Hence it remains to show that $2 | \langle p_1(E_0), [S^4] \rangle$. For this we use the following fact

Lemma 4.24. *The stable quaternionic line bundle $\eta^4 \oplus \Theta^1$ generates $\widetilde{KO^0}(S^4) = \pi_3 SO(5)$.*

Proof. By the definition of η^4 (last section) we see that η^4 results from the clutching construction applied to $A\colon S^3 \to SO(4)$, which is the splitting of $q\colon SO(4) \to S^3$ given in Remark 4.15; i.e. if $a \in S^3$, then $A_a(x) = ax$. Recall $SO(4) \cong SO(3) \times S^3$, hence $\pi_3 SO(4) = \pi_3 SO(3) \oplus \pi_3(S^3)$. Therefore η and φ form a basis for $\pi_3 SO(4)$, where φ is the image of a generator of $\pi_3 SO(3)$.

Now let α generate $\pi_3 SO(5)$ and consider the homotopy exact sequence for the fiber bundle $SO(3) \to SO(5) \to V_{2,5}$, where $V_{2,5}$ is the space of orthonormal 2-frames in $\mathbb{R}^5$

$$\pi_3 SO(3) \to \pi_3 SO(5) \to \pi_3(V_{2,5}) \to \pi_2 SO(3)$$

where $\pi_3 SO(3) \cong \mathbb{Z}$ and $\pi_2 SO(3) = \pi_2(\mathbb{RP}^3) = 0$.

It is easy to see that $V_{2,5}$ is just the sphere bundle of the tangent bundle of S^4 and there is a fiber bundle $S^3 \to V_{2,5} \to S^4$ (Figure 4.2).

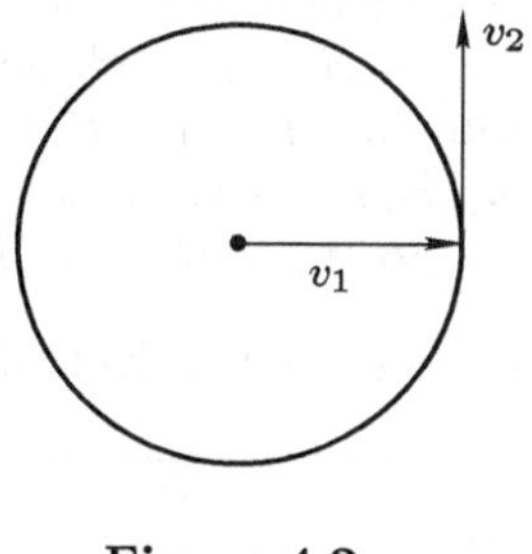

Figure 4.2

By Gysin sequence

$$H^0(S^4) \xrightarrow{\cup e(TS^4)} H^4(S^4) \to H^4(V_{2,5}) \to H^1(S^4) = 0.$$

Therefore $H_3(V_{2,5}) = H^4(V_{2,5}) = \mathbb{Z}_2$. Since $\pi_1(V_{2,5}) = \pi_2(V_{2,5}) = 0$, we have $\pi_3(V_{2,5}) = H_3(V_{2,5}) = \mathbb{Z}_2$. So the image of φ in $\pi_3 SO(5)$ is twice a generator; i.e. the stable bundle $\varphi \oplus \Theta^1 = \pm 2\alpha$. But since $\pi_3 SO(4) \to \pi_3 SO(5)$ is a surjection, the image of η and φ must generate $\pi_3 SO(5)$. Thus $\eta \oplus \Theta^1 = s\alpha$ with s an odd integer. As seen in the last section, the first Pontryagin number $\langle p_1(-), [S^4]\rangle$ is a group homomorphism from $\widetilde{KO}^0(S^4)$ to $\mathbb{Z}$. Therefore

$$-2 = \langle p_1(\eta), [S^4]\rangle = s\langle p_1(\alpha), [S^4]\rangle.$$

This implies that $s = \pm 1$. □

As a consequence, we see that $2|\langle p_1(E_0), [S^4]\rangle$, because

$$\langle p_1(E_0), [S^4]\rangle = \langle p_1(E_0 \oplus \Theta^1), [S^4]\rangle = \langle p_1(t(\eta \oplus \Theta^1)), [S^4]\rangle = -2t.$$

So it remains to verify that $J(\xi \oplus \Theta^1) = 0$. For this we need to change perspective a bit. Instead of TM^4 we should focus on the stable normal bundle ν of M; i.e. the negative of TM^4 in $\widetilde{KO}^0(S^4)$. To be specific, embed M^4 in S^9 and let ν be its normal bundle. Hence $TM \oplus \nu = \Theta^9$. There is a natural degree-one map $\gamma\colon S^9 \to T(\nu) = \mathbb{D}(\nu)/S(\nu)$, where $T(\nu)$ is the Thom complex of ν, and γ is defined by

$$\gamma(x) = \begin{cases} x, & \text{if } x \in \operatorname{int}(\mathbb{D}(\nu)), \\ \infty, & \text{if } x \in S^9 - \operatorname{int}(\mathbb{D}(\nu)). \end{cases}$$

Note that $\nu|_{\mathbb{D}^4}$ and $\nu|_{M-\operatorname{int}\mathbb{D}^4}$ are both trivial since $TM|_{\mathbb{D}^4}$ and $TM|_{M-\operatorname{int}\mathbb{D}^4}$ are trivial. So the clutching construction can be performed to define a bundle map $\overline{f}\colon \nu \to \rho$ covering $f\colon M \to S^4$. Clearly the composite γ and $\overline{f}$ induces a degree-one map $\overline{\gamma}\colon S^9 \to T(\rho)$. We use this to show that $J(\rho) = 0$. Consider the wedge of $\overline{\gamma}$ and the fiber inclusion map $\sigma\colon \mathbb{D}^5/S^4 \to T(\rho)$; i.e. $\overline{\gamma} \vee \sigma\colon S^9 \vee S^5 \to T(\rho)$. By Thom isomorphism $\widetilde{H}_{i+5}(T(\rho)) \cong H_i(S^4)$. Hence it is easy to see that $\overline{\gamma} \vee \sigma$ induces an isomorphism on homology and consequently is a homotopy equivalence by Whitehead Theorem. But $T(\rho) = S^5 \cup_{J(\rho)} \mathbb{D}^9$; hence $J(\rho) = 0$.

Therefore by the above argument $\langle p_1(\nu), [M^4]\rangle$ is divisible by 48. But since the equation $p_1(E_1 \oplus E_2) = p_1(E_1) + p_1(E_2)$ holds in $H^4(M;\mathbb{Q})$, we see that $p_1(TM) = -p_1(\nu)$ in $H^4(M^4;\mathbb{Z})$. So $\langle p_1(TM), [M^4]\rangle$ is divisible by 48. □

Remark 4.25. The reference of this section is Kervaire-Milnor's article [9].

4.5 Proof of Novikov's Theorem

The object of this section is to give a proof of Novikov's Theorem 4.5; i.e.

Theorem 4.26. *Let $E^m \to S^n$ be an m-dimensional vector bundle over S^n, where $m > n$. If E^m is topologically trivial, then it is a trivial vector bundle; i.e. $E^m \cong \Theta^m$.*

Our proof will depend on the following two results whose proofs are (in turn) consequences of the handlebody arguments used in proving the h- (and s-) Cobordism Theorems. After using them to prove Novikov's result, we will briefly sketch their proofs.

Theorem 4.27 (splitting theorem). *Let W^n be a smooth manifold ($n \geq 6$) which is* homeomorphic *to $M^{n-1} \times \mathbb{R}$, where M^{n-1} is a closed smooth manifold. If $\pi_1(M)$ is free abelian, then W^n is diffeomorphic to $N^{n-1} \times \mathbb{R}$ for some closed smooth manifold N^{n-1}.*

Remark 4.28. When $\pi_1(M) = 1$, this result is due to W. Browder (1965) and in the general case due to Novikov (1966) and independently Siebenmann (1965 Ph. D. thesis).

Theorem 4.29 (fibering theorem). *Let M^m and K^{m-1} be closed smooth manifolds ($m \geq 6$) such that M^m is homotopy equivalent to $K \times S^1$ (via f). If $\pi_1(K)$ is free abelian, then M^n smoothly fibers over the circle S^1 with fiber N^{m-1} homotopy equivalent to K^{m-1} (via φ) such that $N \subset M \xrightarrow{f} K \times S^1$ is homotopic to $N \xrightarrow{\varphi} K \subset K \times S^1$.*

Remark 4.30. When $\pi_1(K) = 1$, this result is due to Browder and Levine (1966) and in the general case due to Farrell (1967 Ph. D. thesis).

Remark 4.31. Novikov's proof of Theorem 4.26 (1966) did not use Theorem 4.29; but instead a more complicated version of Theorem 4.27 and some other arguments related to what eventually has become known as "Novikov Conjecture". But the use of Theorem 4.29 makes the proof of Novikov's Theorem more transparent.

Anchor Rings

Definition 4.32. *An anchor ring $\mathfrak{a}$ is an open subset of $\mathbb{R}^n$ diffeomorphic to $T^{n-1} \times \mathbb{R}$, where T^{n-1} is the $(n-1)$-dimensional torus $\underbrace{S^1 \times \cdots \times S^1}_{(n-1) \text{ factors}}$.*

We first note that anchor rings exist in each $\mathbb{R}^n$. This is proved by induction on n starting with $n = 2$ where $\mathfrak{a}$ is shown in Figure 4.3.

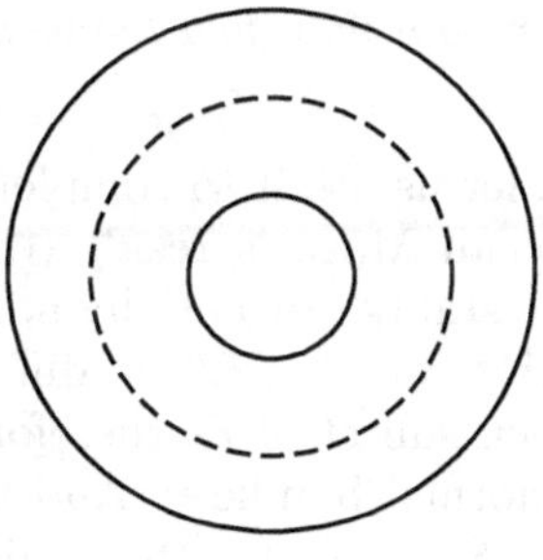

Figure 4.3

Given an anchor ring $\mathfrak{a} \subset \mathbb{R}^n$, this gives an embedding of T^{n-1}, $T^{n-1} \subset \mathbb{R}^n \subset \mathbb{R}^{n+1}$, with trivial normal bundle ν. Let $\hat{\mathfrak{a}} = \nu - \{\text{zero} - \text{section}\}$. Then $\hat{\mathfrak{a}} \subset \mathbb{R}^{n+1}$ and $\hat{\mathfrak{a}} = T^{n-1} \times (\mathbb{R}^2 - 0) = T^n \times \mathbb{R}$. This proves the existence in general.

We first prove Theorem 4.26 under the extra assumption that $n \geq 5$. And then give special arguments for the remaining 4 cases (of which only the case $n = 4$ is difficult).

Let $f\colon E \to S^n \times \mathbb{R}^m$ be a homeomorphism and $\mathfrak{a} \subset \mathbb{R}^m$ be an anchor ring. Let $W^{n+m} = f^{-1}(S^n \times \mathfrak{a})$ which is a smooth $(n+m)$-dimensional manifold since *it is an open subset of E which is smooth.* Hence $f\colon W \to (S^n \times T^{m-1}) \times \mathbb{R}$ is a homeomorphism. And $\pi_1(S^n \times T^{m-1}) = \mathbb{Z}^{m-1}$. Therefore by Theorem 4.27, W is diffeomorphic to $N_1 \times \mathbb{R}$ where N_1 is a closed smooth manifold homotopy equivalent (but perhaps *not* homeomorphic) to $S^n \times T^{m-1} = (S^n \times T^{m-2}) \times S^1$. Now applying Theorem 4.29, N_1 fibers over S^1, $N_1 \to S^1$, with fiber N_2 homotopy equivalent to $S^n \times T^{m-2} = (S^n \times T^{m-3}) \times S^1$. Applying Theorem 4.29 again shows N_2 fibers over S^1, $N_2 \to S^1$, with fiber N_3 homotopy equivalent to $S^n \times T^{m-3} = (S^n \times T^{m-4}) \times S^1$.

Continuing in this way gives a sequence of closed smooth manifolds

$$N_m \subset N_{m-1} \subset N_{m-2} \subset \cdots \subset N_2 \subset N_1 \subset E$$

where each N_i has codimension-1 in N_{i-1} (as does N_1 in E) and each has trivial normal bundle in the next. Hence the normal bundle of N_m in E is Θ^m; in particular $N_m \times \mathbb{R}^m$ is an open subset of E. But N_i (by construction) is homotopy equivalent to $S^n \times T^{m-i}$ via φ_i, such that

$$N_i \xrightarrow{\varphi_i} S^n \times T^{m-i} \subset S^n \times T^{m-1} \times \mathbb{R} \subset S^n \times \mathbb{R}^m$$

is homotopic to

$$N_i \subset E \xrightarrow{f} S^n \times \mathbb{R}^m,$$

and (in particular) N_m is homotopy equivalent to S^n via φ_m and hence homeomorphic to S^n such that $N_m \xrightarrow{\varphi_m} S^n \subset S^n \times \mathbb{R}^m$ is homotopic to $N_m \subset E \xrightarrow{f} S^n \times \mathbb{R}^m$. And $N_m \subset E$ is a homotopy equivalence. To proceed further we must first prove the following

Lemma 4.33. *Let Σ^n be a smooth manifold homeomorphic to S^n. Then $T\Sigma^n \oplus \Theta^1 = \Theta^{n+1}$.*

Before proving this result, let us use it to complete the proof of Theorem 4.26 (in the case $n \geq 5$). Now recall Mazur's result ([12]) that two closed smooth manifolds M_1 and M_2 that are stably tangentially homotopy equivalent are stably diffeomorphic. Hence $N_m \times \mathbb{R}^m = \Sigma^n \times \mathbb{R}^m$ is diffeomorphic to $S^n \times \mathbb{R}^m$. But recall that $N_m \times \mathbb{R}^m$ is an open subset of E, therefore S^n is smoothly embedded in E (say, by τ) with trivial normal bundle ν. And this embedding is homotopic to the 0-section embedding $\sigma\colon S^n \to E$. By the Whitney Embedding Theorem σ is isotopic to τ. And hence the normal bundle to σ is also trivial. But we have seen that $TE|_{\sigma(S^n)} = E \oplus TS^n$. Hence E is the normal bundle to σ. Therefore $E \cong \Theta^m$. This completes the proof of Theorem 4.26 when $n \geq 5$.

Proof of Lemma 4.33. Recall by Bott periodicity that all stable bundles over S^n are trivial if $n \equiv 3, 5, 6, 7 \pmod 8$. So we need only consider the four cases $n \equiv 0, 1, 2, 4 \pmod 8$.

To handle the case $n \equiv 0$ or $4 \pmod 8$ (i.e. $n = 4s$ for $s \in \mathbb{Z}^+$) we use the Hirzebruch index formula which says that

$$0 = I(S^{4s}) = I(\Sigma^{4s}) = \langle a_s p_s(\Sigma^{4s}), [\Sigma^{4s}] \rangle$$

where $a_s \in \mathbb{Q} - \{0\}$. This implies that $p_s(\Sigma^{4s}) = 0$. Hence $T\Sigma^{4s}$ has finite order in $\widetilde{KO^0}(\Sigma^{4s}) = \widetilde{KO^0}(S^{4s}) = \mathbb{Z}$. Therefore $T\Sigma^{4s} \oplus \Theta^1 = \Theta^{4s+1}$. (In fact,

$$a_s = \frac{2^{2s}(2^{2s-1} - 1)B_s}{(2s)!}$$

where B_s is the s-th Bernoulli number, which is given by

$$\frac{z}{e^z - 1} = 1 - \frac{z}{2} + \frac{B_1}{2!} z^2 - \frac{B_2}{4!} z^4 + \frac{B_3}{6!} z^6 - \cdots)$$

To handle the case $n \equiv 1, 2 \pmod 8$ we use a fundamental result due to Frank Adams that $J\colon \pi_{n-1}(O(m)) \to \pi_{n+m-1}(S^m)$ is monic when $n \equiv 1, 2 \pmod 8$ and $m > n$. Next observe that $T\Sigma^n \oplus \Theta^1 = \Theta^{n+1}$ if and only if $\nu = \Theta^{n+1}$ where ν is the normal bundle of an embedding of Σ^n in S^{2n+1}. (Recall ν represents $-T\Sigma^n$ in $\widetilde{KO^0}(\Sigma^n)$.) Now use the Whitney Embedding Theorem to construct an embedding $\Sigma^n \subset S^n \times \mathbb{D}^{n+1} \subset S^{2n+1}$. Hence $\mathbb{D}(\nu) \subset S^n \times \mathbb{D}^{n+1}$ and the "complementary region" W^{2n+1} is an h-cobordism which is trivial since $\pi_1(W) = 0$. Therefore

$$\mathbb{D}(\nu)/S(\nu) = S^n \times \mathbb{D}^{n+1}/S^n \times S^n \simeq S^{2n+1} \vee S^{n+1}.$$

Hence (as seen in the previous section) ν is fiber homotopically trivial; i.e. $J(\nu) = 0$. Consequently $\nu = 0$ in $\widetilde{KO^0}(\Sigma^n)$ completing the proof of Lemma 4.33. □

Now we deal with the remaining cases of Theorem 4.26.

For $n = 3$, Bott showed that (this is overkill) $\widetilde{KO^0}(S^3) = 0$. Thus $E^m = \Theta^m$. For $n = 1$ or 2, $J\colon \widetilde{KO^0}(S^n) \to \pi_{m+n}(S^n)$ is monic (by Adams, again overkill). So it suffices (in these two cases) to show that $T(E) \simeq S^{m+n} \vee S^m$. But this is clear since E is homeomorphic to $S^n \times \mathbb{R}^m$ and $T(E) = E \cup \{\infty\}$.

So it remains to focus on the special case where $n = 4$. Since $\widetilde{KO^0}(S^4) = \mathbb{Z}$ which is torsion free, it is sufficient to show that $p_1(E) = 0$. Unfortunately our argument for the case $n \geq 5$ doesn't quite work since to get the manifold $\Sigma^n = N_m$ we would need to use Theorem 4.29 on a 5-dimensional manifold (where the result is perhaps false). To get around this, we pull back the bundle E^m via the projection $\pi\colon S^4 \times S^1 \to S^4$ to get a stable bundle $\mathscr{E}^m$ over $S^4 \times S^1$ which is topologically trivial. (For this we should replace E^m by $E^m \oplus \Theta^1$ if $m = 5$ so $\mathscr{E}$ is stable.) Now we will imitate the argument in the case $n = 5$, and will eventually show that

Lemma 4.34. *Let $\sigma\colon S^4 \to S^4 \times S^1$ be the obvious inclusion, then $\langle p_1(\mathscr{E}), \sigma_*[S^4] \rangle = 0$.*

Assuming momentarily this lemma, we complete the proof for $n = 4$. Note that

$$\begin{aligned}\langle p_1(\mathscr{E}), \sigma_*([S^4])\rangle &= \langle \sigma^* p_1(\mathscr{E}), [S^4]\rangle \\ &= \langle p_1(\sigma^*(\mathscr{E})), [S^4]\rangle \\ &= \langle p_1(\sigma^*\pi^*(E)), [S^4]\rangle \\ &= \langle p_1(E), [S^4]\rangle .\end{aligned}$$

Therefore Lemma 4.34 implies that $p_1(E) = 0$ and we are done.

Hence it remains to prove Lemma 4.34. Using the argument used in the case $n \geq 5$, we construct a smooth closed 5-manifold M^5 and a smooth embedding $f\colon M^5 \times \mathbb{R}^m \hookrightarrow \mathscr{E}$ which is a homotopy equivalence. Let $q\colon \mathscr{E} \to S^4 \times S^1$ be the projection, $i\colon M^5 \to M^5 \times \mathbb{R}^m$ be the obvious inclusion, $\varphi\colon M \to S^4 \times S^1$ be the homotopy equivalence $q \circ f \circ i$. We have the following commutative diagram

$$\begin{array}{ccc} M \times \mathbb{R}^m & \xrightarrow[\simeq]{f} & \mathscr{E}^m \\ {\scriptstyle i}\uparrow & & \downarrow{\scriptstyle q} \\ M & \xrightarrow[\simeq]{\varphi} & S^4 \times S^1 \end{array}$$

Now $T\mathscr{E} = q^*(T(S^4 \times S^1) \oplus \mathscr{E})$. Hence $T(M^5 \times \mathbb{R}^m) = f^*q^*(T(S^4 \times S^1) \oplus \mathscr{E})$. Thus $TM^5 \oplus \Theta^m = i^*f^*q^*(T(S^4 \times S^1) \oplus \mathscr{E})$ and $TM^5 = \varphi^*(\mathscr{E})$ in $\widetilde{KO^0}(M)$. Let $\psi\colon S^4 \times S^1 \to M$ be the homotopy inverse of φ, then $\psi^*(TM^5) = \mathscr{E}$ in $\widetilde{KO^0}(S^4 \times S^1)$. Therefore

$$\begin{aligned}\langle p_1(\mathscr{E}), \sigma_*([S^4])\rangle &= \langle p_1\psi^*(TM^5), \sigma_*([S^4])\rangle \\ &= \langle \psi^* p_1(M), \sigma_*([S^4])\rangle \\ &= \langle p_1(M), \psi_*\sigma_*([S^4])\rangle .\end{aligned}$$

Then the proof of Lemma 4.34 is reduced to the following:

Proposition 4.35. *Let $\varphi\colon M^5 \to S^1 \times S^4$ be a homotopy equivalence. Then $p_1(M^5) = 0$ in $H^4(M;\mathbb{Q})$.*

Proof. Consider the continuous map $g\colon M^5 \to S^1$ which is the composite of φ and the projection $S^5 \times S^1 \to S^1$. Wiggle it a bit so it is transverse to $1 \in S^1 \subset \mathbb{C}$. Then $F^4 = g^{-1}(1)$ is a closed smooth 4-manifold with trivial normal bundle in M, and such that $\varphi_*[F^4] = [S^4] \in H_4(S^4 \times S^1)$. We can arrange that F^4 is connected, although this is not essential for the following arguments.

Here is an argument to show this. Let $F_1, F_2, \cdots, F_m$ be the connected components of F. $M - F_i$ has either one or two components. If $M - F_i$ has 2 components A and B, then $F_i = \partial A$ and $[F_i] = 0$ in $H_4(M)$. Since $[F]$ which represents a generator of $H_4(M;\mathbb{Z})$ is $\sum_i [F_i]$, some $[F_i] \neq 0$ and hence $M - F_i$ is connected. Renumber things so F_1 has this property. Let α be a short simple arc with $\alpha(0)$, $\alpha(1)$ on left, right side of F_1 with $F_1 \cap \alpha = \mathrm{pt}$ (Figure 4.4). Let γ be an arc in $M - F_1$ connecting $\alpha(0)$ to $\alpha(1)$; i.e. $\gamma(0) = \alpha(1)$ and $\gamma(1) = \alpha(0)$. Then joining endpoints produces a map $S^1 \to M^5$ which is transverse to F_1, and $F_1 \cap S^1 = \mathrm{pt}$. Therefore $[F_1]$ is a generator of $H_4(M;\mathbb{Z})$.

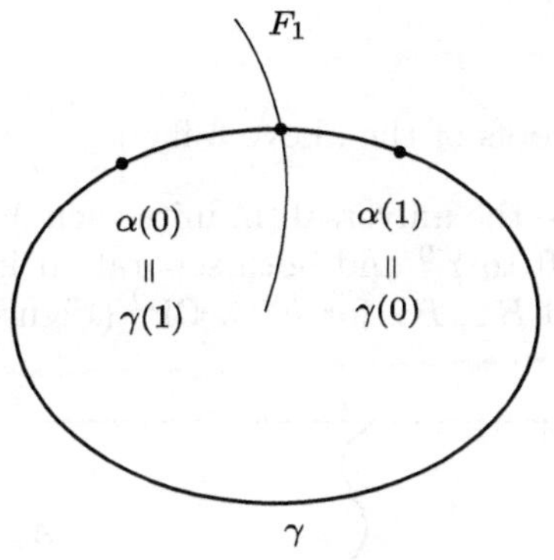

Figure 4.4

Let $j\colon F \to M$ be the inclusion, then

$$\langle p_1(M), j_*[F]\rangle = \langle j^* p_1(M), [F]\rangle = \langle p_1(F^4), [F]\rangle = 3I(F).$$

Therefore Proposition 4.35 is equivalent to

Lemma 4.36. $I(F^4) = 0$.

Proof. This would follow directly from Theorem 4.29 if it were true for 5-manifolds; which unfortunately is not.

To get around this we proceed as follows: consider

$$f \times \mathrm{id}\colon M^5 \times \mathbb{C}\mathrm{P}^2 \to (S^1 \times S^4) \times \mathbb{C}\mathrm{P}^2.$$

This is still a homotopy equivalence and now the dimension is bigger than 5, so the fibering theorem, Theorem 4.29 applies to yield a smooth submanifold $N^8 \subset M^5 \times \mathbb{C}\mathrm{P}^2$ with $N^8 \simeq S^4 \times \mathbb{C}\mathrm{P}^2$.

We now use the following 3 facts to complete the proof of this lemma, and then verify these facts.

Fact 1. N^8 is oriented cobordant to $F^4 \times \mathbb{C}\mathrm{P}^2$.

Fact 2. Oriented cobordant manifolds have the same index.

Fact 3. Let K_1^{4n} and K_2^{4m} be a pair of closed oriented manifolds, then

$$I(K_1 \times K_2) = I(K_1) \times I(K_2)$$

.

Using these facts, we have that

$$I(N^8) = I(F^4 \times \mathbb{C}\mathrm{P}^2) = I(F^4) \times I(\mathbb{C}\mathrm{P}^2) = I(F^4).$$

Now recall that N^8 is homotopy equivalent to $S^4 \times \mathbb{C}\mathrm{P}^2$, therefore

$$I(N^8) = I(S^4) \times I(\mathbb{C}\mathrm{P}^2) = 0.$$

Consequently, $I(F^4) = 0$, verifying Lemma 4.36 and hence Proposition 4.35. □

□

Now we turn to the proofs of the above 3 facts.

Proof of Fact 1. Let Y^9 be the universal (infinite cyclic) cover of $M^5 \times \mathbb{C}P^2$. Then N^8 and $F^4 \times \mathbb{C}P^2$ both lift to Y^9 and both separate it in two unbounded components, A_-, A_+ for N^8 and B_-, B_+ for $F^4 \times \mathbb{C}P^2$ (Figure 4.5).

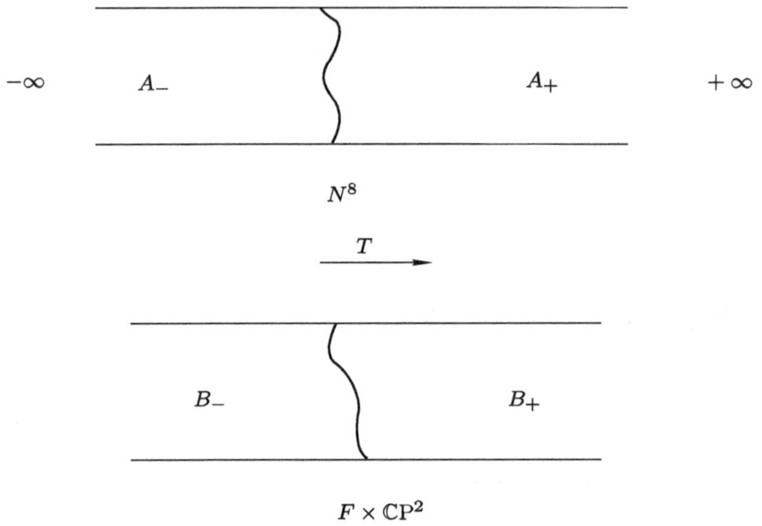

Figure 4.5

Since positive powers of the generator T of the infinite cyclic group of deck transformations of the covering space $Y \to M \times \mathbb{C}P^2$ moves points in Y towards $+\infty$, we can pick our lift of $F \times \mathbb{C}P^2$ so that $F^4 \times \mathbb{C}P^2 \subset \text{int}(A_+)$ (Figure 4.6).

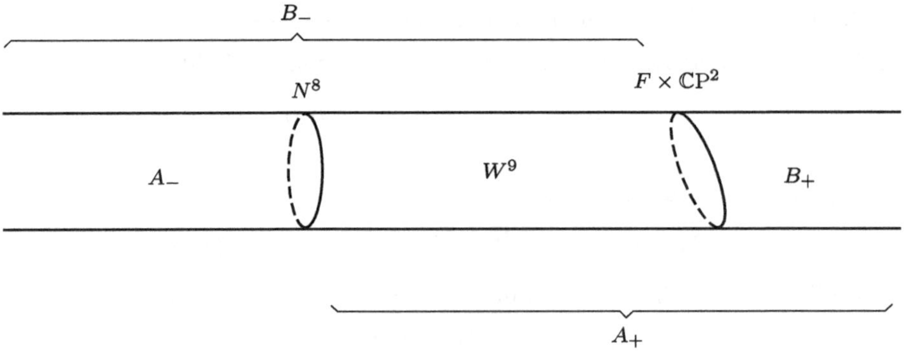

Figure 4.6

Let $W^9 = A_+ \cap B_-$. Then W^9 is the desired cobordism between N^8 and $F^4 \times \mathbb{C}\mathrm{P}^2$. □

Proof of Fact 2. Let W^{4n+1} be an oriented cobordism between M^{4n} and N^{4n}; i.e. $\partial W = M \cup (-N)$ (Figure 4.7).

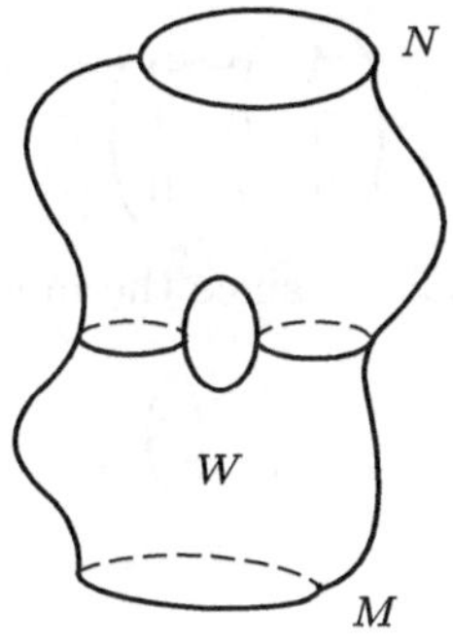

Figure 4.7

Let $\sigma\colon \partial W \to W$ denote the inclusion. Then $T(\partial W) \oplus \Theta^1 = \sigma^*(TW)$ since $\nu_\sigma = \Theta^1$. Therefore $p_i(\partial W) = p_i(\sigma^*(TW)) = \sigma^*(p_i W)$. Hence $l_n(\partial W) = \sigma^*(l_n W)$, and

$$I(\partial W) = \langle l_n(\partial W), [\partial W]\rangle = \langle \sigma^*(l_n(W), [\partial W]\rangle = \langle l_n(W), \sigma_*[\partial W]\rangle.$$

Clearly $I(\partial W) = I(M) - I(N)$ and $\sigma_*[\partial W] = 0$ in $H_{4n}(W)$. This shows $I(M) = I(N)$. □

Remark 4.37. There is an elementary (but longer) proof of Fact 2 which does not use the index formula. In fact the original proof of the index formula depended on this elementary proof.

Proof of Fact 3. We only need an important special case of this formula, namely where $K_2 = \mathbb{C}\mathrm{P}^2$, hence abbreviate K_1 by K^{4n} and we proceed to show that

$$I(K^{4n} \times \mathbb{C}\mathrm{P}^2) = I(K).$$

For this let $a_1, \cdots, a_s$ be a basis for $H^{2n}(K;\mathbb{R})$ and $b_1, \cdots, b_t$ be a basis for $H^{2n+2}(K;\mathbb{R})$. Let $c_1, \cdots, c_t$ be its Poincaré dual basis for $H^{2n-2}(K;\mathbb{R})$; i.e.

$$\langle b_i \cup c_j, [K]\rangle = \begin{cases} 0, & \text{if } i \neq j, \\ 1, & \text{if } i = j. \end{cases}$$

Let A be the symmetric matrix $A_{ij} = \langle a_i \cup a_j, [K]\rangle$. Let $\alpha = 1 \in H^0(\mathbb{C}\mathrm{P}^2;\mathbb{R})$, $\gamma \in H^4(\mathbb{C}\mathrm{P}^2;\mathbb{R})$ be the element such that $\langle \gamma, [\mathbb{C}\mathrm{P}^2]\rangle = 1$ and β generate $H^2(\mathbb{C}\mathrm{P}^2;\mathbb{R})$ such that $\langle \beta \cup \beta, [\mathbb{C}\mathrm{P}^2]\rangle = 1$. Now

$$H^{2n+2}(K \times \mathbb{C}\mathrm{P}^2)$$
$$= (H^{2n}(K) \otimes H^2(\mathbb{C}\mathrm{P}^2)) \oplus (H^{2n+2}(K) \otimes H^0(\mathbb{C}\mathrm{P}^2)) \oplus (H^{2n-2}(K) \otimes H^4(\mathbb{C}\mathrm{P}^2))$$

has basis $a_1 \otimes \beta, \cdots, a_s \otimes \beta$, $b_1 \otimes \alpha, \cdots, b_t \otimes \alpha$, $c_1 \otimes \gamma, \cdots, c_t \otimes \gamma$ (for the 3 summands respectively).

In terms of this basis the matrix representing the intersection form of $K \times \mathbb{C}\mathrm{P}^2$ is

$$\begin{pmatrix} A & 0 & 0 \\ 0 & 0 & I \\ 0 & I & 0 \end{pmatrix}.$$

Hence $I(K \times \mathbb{C}\mathrm{P}^2) = I(A) = I(K^{4n})$ since the index of the block matrix

$$\begin{pmatrix} 0 & I \\ I & 0 \end{pmatrix}$$

is 0. □

4.6 Novikov Conjecture

Novikov's proof in case $n = 4$ of his Theorem 4.26 depends on the following "homotopy invariance" result he proved a few months earlier in 1965.

To describe this result we need first recall the Hirzebruch Index Theorem. Remember $H^*(BO;\mathbb{Q}) = \mathbb{Q}[p_1, p_2, \cdots, p_n, \cdots]$ where $p_i \in H^{4i}(BO;\mathbb{Z})$. Hirzebruch constructed different polynomial generators $l_i \in H^{4i}(BO;\mathbb{Q})$ such that $H^*(BO;\mathbb{Q}) = \mathbb{Q}[l_1, l_2, \cdots, l_n, \cdots]$. For example, $l_1 = \frac{1}{3}p_1$, $l_2 = \frac{1}{45}(7p_2 - p_1^2)$, $\cdots$. Although these are *not* integral classes they have the following important property.

Hirzebruch Index Theorem 4.38. *For every closed smooth oriented manifold M^{4n},*

$$I(M^{4n}) = \langle l_n(M), [M] \rangle.$$

Since $I(M)$ is clearly a homotopy invariant, Hirzebruch Index Theorem has the following corollary.

Corollary 4.39. *$l_n(M^{4n})$ is a homotopy invariant; i.e. if $f\colon N^{4n} \to M^{4n}$ is a homotopy equivalence between closed smooth manifolds, then $f^*(l_n(M^{4n})) = l_n(N^{4n})$.*

Novikov generalized this to the following

Novikov Homotopy Invariance Theorem 4.40. *Let $f\colon N^{4n+1} \to M^{4n+1}$ be a homotopy equivalence between closed smooth manifolds and $a \in H^1(M)$. Then $f^*(l_n(M) \cup a) = l_n(N) \cup f^*(a)$.*

Consequently, by Poincaré duality, $f^*(l_n(M^{4n+1})) = l_n(N^{4n+1})$.

He then conjectured in 1966:

Original Novikov Conjecture 4.41. *Let $f\colon M^{4n+k} \to N^{4n+k}$ be a homotopy equivalence between closed (oriented) smooth manifolds and $a_1, a_2, \cdots, a_k \in H^1(N)$ be any k elements. Then $f^*(l_n(N) \cup a_1 \cup \cdots \cup a_k) = l_n(M) \cup f^*(a_1) \cup \cdots \cup f^*(a_k)$.*

Remark 4.42. This conjecture does not necessarily implies $f^*(l_n(N)) = l_n(M)$ since $H^k(M)$ needs not to be generated by 1-dimensional cohomology classes.

Remark 4.43. Hsiang and Farrell in 1967 proved this conjecture using a generalization of Theorem 4.29.

In the meantime Novikov in 1968 extended his conjecture to the following:

Novikov Conjecture 4.44. *Let $f\colon M^{4n+k} \to N^{4n+k}$ be a homotopy equivalence between closed smooth manifolds and $a \in H^k(\pi_1(N))$. Then*

$$f^*(l_n(N) \cup \varphi^*(a)) = l_n(M) \cup (\varphi \circ f)^*(a)$$

where $\varphi\colon N \to K(\pi_1(N), 1)$ classifies the fundamental group of N.

Remark 4.45. This is the conjecture which is usually meant when people refer to "Novikov Conjecture".

Remark 4.46. Recall $[X, K(\pi, 1)]$ is in bijective correspondence with $\hom(\pi_1(X), \pi)$ via $f\colon X \to K(\pi, 1)$ corresponds to $f_*\colon \pi_1(X) \to \pi_1(K(\pi, 1)) = \pi$.

Remark 4.47. If N is aspherical, then the Novikov Conjecture implies $f^*(p_i(N)) = p_i(M) \in H^{4i}(M; \mathbb{Q})$ for any homotopy equivalence $f\colon M \to N$.

Proof. For $4n \leq \dim M$, write $4n+k = \dim M$ and note that $\varphi\colon N \to K(\pi_1(N), 1)$ is the identity. Therefore for every $a \in H^k(N; \mathbb{Q})$

$$f^*(l_n(N)) \cup f^*(a) = f^*(l_n(N) \cup a) = l_n(M) \cup f^*(a),$$

hence Poincaré duality (if N is non-orientable, extra argument is needed) shows that $f^*(l_n(N)) = l_n(M)$. But the Pontryagin classes p_i are polynomials in l_n. Therefore $f^*(p_i(N)) = p_i(M)$. □

Remark 4.48. If N is aspherical and Borel Conjecture is true for N, then the Novikov Conjecture is true for N.

Proof. Let $f\colon M \to N$ be a homotopy equivalence. If the Borel Conjecture is true for N, f is homotopic to a homeomorphism $g\colon M \to N$ and by Novikov's Theorem 4.8

$$f^*(p_i(N)) = g^*(p_i(N)) = p_i(M).$$

But $l_n(N)$ is a polynomial in $p_i(N)$, therefore $f^*(l_n(N)) = l_n(M)$ and

$$f^*(l_n(N) \cup \varphi^*(a)) = f^*(l_n(N)) \cup f^*\varphi^*(a) = l_n(M) \cup f^*\varphi^*(a).$$ □

5

Exotic Spheres and Surgery

5.1 Plumbing

Consider the following picture

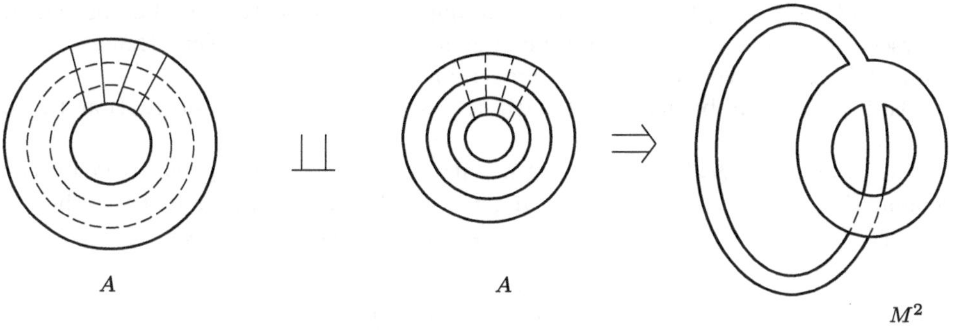

Figure 5.1

where we have glued together two copies of an annulus A via identifying the two cross hatched "discs" by interchanging the solid and dashed parallel lines resulting in a 2-dimensional manifold M^2. Also note that ∂M is connected; i.e. $\partial M = S^1$ and that A is the disc bundle of the tangent bundle TS^1 of the circle since S^1 is parallelizable. M^2 is called two copies of TS^1 plumbed together.

And we can do this construction for any sphere S^n (in fact any closed manifold N^n) to get a "smooth" manifold M^{2n} with boundary (corners must be rounded as usual). And M^{2n} is called two copies of TS^n plumbed together. (Figure 5.2, $\mathbb{D}S^n \sqcup \mathbb{D}S^n \Rightarrow M^{2n}$, here $\mathbb{D}^n \times \mathbb{D}^n = \mathbb{D}S^n|_{\text{northern hemisphere}}$).

Lemma 5.1. *∂M^{2n} is homeomorphic to S^{2n-1} if and only if n is odd.*

Proof. First note that M^{2n} is homotopy equivalent to $S^n \vee S^n$ and consider the exact sequence in homology for the pair $(M^{2n}, \partial M^{2n})$; in particular the map

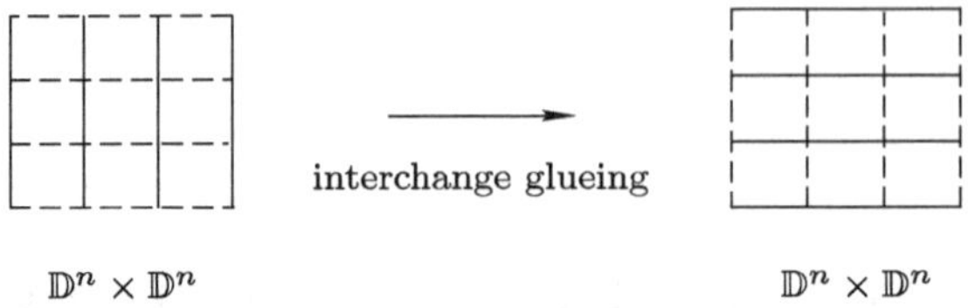

Figure 5.2

$\tau\colon H_n(M^{2n}) \to H_n(M^{2n}, \partial M^{2n})$. By Lefschetz duality $H_n(M^{2n}, \partial M^{2n})$ is identified with $H^n(M^{2n})$ and the universal coefficient theorem gives a natural map $\varphi\colon H^n(M^{2n}) \to \hom(H_n(M^{2n}), \mathbb{Z})$ which is an isomorphism since $\widetilde{H}_i(M^{2n}) = 0$ for $i < n$. Now the composite map

$$H_n(M^{2n}) \xrightarrow{\tau} H_n(M^{2n}, \partial M^{2n}) = H^n(M^{2n}) \xrightarrow{\varphi} \hom(H_n(M^{2n}), \mathbb{Z})$$

can be studied geometrically as follows. By the Hurewicz Theorem and Whitney Embedding Theorem, a homology class $x \in H_n(M^{2n})$ can be realized by an embedding $x\colon S^n \subset M^{2n}$. Hence we define an intersection bilinear form $- \wedge -\colon H_n(M) \times H_n(M) \to \mathbb{Z}$ by defining $x \wedge y$ to be the number of signed intersection points of $x \cap y$ after making y transverse to x. Note that $x \wedge y$ is symmetric if n is even and skew-symmetric if n is odd.

And $\varphi \circ \tau$ can be identified with the adjoint linear transformation to $\wedge$; i.e. $\varphi(\tau(x))(y) = x \wedge y$.

Now obviously $H_n(M^{2n}) = \mathbb{Z} \oplus \mathbb{Z}$ with a natural basis e_1, e_2 given by the "cores" (i.e. 0-sections) of the two tangent bundles that are plumbed together to form M^{2n}. The 2×2 matrix representing $\wedge$ in terms of this basis is

$$\begin{pmatrix} e_1 \wedge e_1 & \pm 1 \\ 1 & e_2 \wedge e_2 \end{pmatrix}.$$

But

$$e_1 \wedge e_1 = e_2 \wedge e_2 = \chi(S^n) = \begin{cases} 0, & \text{if } n \text{ is odd,} \\ 2, & \text{if } n \text{ is even.} \end{cases}$$

Therefore $- \wedge -$ is represented by

$$\begin{pmatrix} 0 & -1 \\ 1 & 0 \end{pmatrix} \text{ if } n \text{ is odd,}$$

and

$$\begin{pmatrix} 2 & 1 \\ 1 & 2 \end{pmatrix} \text{ if } n \text{ is even.}$$

Hence the adjoint to $- \wedge -$ is an isomorphism if and only if n is odd. (The determinant of the matrix is 1 when n is odd and 3 when n is even.) Therefore τ is an isomorphism when n is odd and $\operatorname{coker} \tau = \mathbb{Z}_3$ when n is even. Now let's use this information to analyze the exact sequence for $(M, \partial M)$

$$H_{n+1}(M, \partial M) \to H_n(\partial M) \to H_n(M) \xrightarrow{\mathcal{I}} H_n(M, \partial M) \to H_{n-1}(M) = 0.$$

Note that $H_{n+1}(M, \partial M) \cong H^{n-1}(M) = 0$. Thus if n is odd, $H_i(\partial M) = 0$ for $i = n$ and $i = n - 1$; if n is even, $H_{n-1}(\partial M) = \mathbb{Z}_3$. In particular ∂M^{2n} is not homeomorphic to S^{2n-1} when n is even.

Now when n is odd and $i < n - 1$, the exact sequence for $(M, \partial M)$ becomes

$$H_{i+1}(M, \partial M) \to H_i(\partial M) \to H_i(M)$$

where $H_{i+1}(M, \partial M) \cong H^{2n-(i+1)}(M) = 0$, and $H_i(M) = 0$ for $i > 0$ and $H_0(M) = \mathbb{Z}$. Therefore ∂M^{2n} is a homology sphere and can be shown to be simply-connected (when $n \neq 1$). Hence ∂M^{2n} is homeomorphic to S^{2n-1} by the h-Cobordism Theorem of Smale. □

See Kervaire-Milnor paper [10] for more details.

Hence we can form a closed topological manifold $K^{4n+2} = M^{4n+2}/\partial M$; i.e. by attaching $\mathbb{D}^{4n+2}$ to M^{4n+2} along their common boundary $\partial M^{4n+2} = S^{4n+1}$. We call these manifolds K^{4n+2} Kervaire manifolds in his honor. He showed (1960) that K^{10} *does not* support a differential structure; in fact K^{10} is not homotopy equivalent to a closed smooth manifold. (Although since then it has been shown that the Hurewicz Conjecture is true for K^{4n+2}; i.e. any closed manifold homotopy equivalent to K^{4n+2} is homeomorphic to it.) And hence ∂M^{10} is a smooth manifold homeomorphic but *not* diffeomorphic to S^9. Kervaire and Milnor (1963) also conjectured that K^{2m} is not smoothable for every odd integer $m \neq 1, 3, 7$ (where it is smoothable) and Browder (1969) verified this conjecture for all $m \neq 2^s - 1$. But it was shown that K^{30} and K^{62} are also smoothable. So the conjecture was modified to say that K^{2m} is *not smoothable* except for finitely many odd integers m. In 2009 this conjecture was verified by Hill, Hopkins and Ravenel. In fact there is only one unsettled case K^{126}.

Remark 5.2. When the Kervaire manifold K^{4n+2} is not smoothable, then the odd dimensional manifold $K^{4n+2} \times S^1$ is also *not smoothable*. In fact all the closed manifolds M^{m+10} homotopy equivalent to $K^{10} \times T^m$ $(m \geq 0)$ are *not smoothable*. Here is the argument. If $M^{m+10} \simeq K^{10} \times T^m = (K^{10} \times T^{m-1}) \times S^1$ is a smooth manifold, then by fibering Theorem 4.29 it smoothly fibers over a circle with smooth fiber F_1 homotopy equivalent to $K^{10} \times T^{m-1}$, which also smoothly fibers over a circle with smooth fiber F_2 homotopy equivalent to $K^{10} \times T^{m-2}$. Continuing in this way m-steps, we finally get a closed smooth manifold F_m homotopy equivalent to K^{10}. But Kervaire showed this was impossible. In particular in every dimension ≥ 10 there exists a closed topological manifold which is *not* smoothable. In fact the following is true.

Proposition 5.3. *$K^{10} \times N^k$ is not smoothable for any manifold N^k.*

Proof. Suppose that $K^{10} \times N^k$ has a smooth structure. Since $K^{10} \times \mathbb{R}^k \subset K^{10} \times N^k$ is an open subset, it has a smooth structure. Recall there is an anchor ring $T^{k-1} \times \mathbb{R} \subset \mathbb{R}^k$ as an open subset, hence $K^{10} \times T^{k-1} \times \mathbb{R}$ has a smooth structure. But by the Novikov-Siebenmann Theorem (splitting Theorem 4.27) $K^{10} \times T^{k-1} \times \mathbb{R}$

is diffeomorphic to $M \times \mathbb{R}$ where M is a closed smooth manifold. M is clearly homotopy equivalent to $K^{10} \times T^{k-1}$. Then we get a contradiction since we have shown in Remark 5.2 that any closed manifold homotopy equivalent to $K^{10} \times T^k$ cannot support a smooth structure. □

Using much more advanced techniques which build on the splitting Theorem 4.27 and fibering Theorem 4.29, Kirby and Siebenmann [11] proved the following

Theorem 5.4. *Let M^m be a topological manifold of dimension $m \geq 5$. Then $M^m \times \mathbb{R}^k$ is smoothable if and only if M^m is smoothable.*

Corollary 5.5. *If M^m is not smoothable, then $M^m \times N^k$ is not smoothable.*

Proof. If $M \times N$ is smoothable, then so is $M \times \mathbb{R}^k$ since it is an open submanifold of $M \times N$. By the above theorem, M is smoothable, which is a contradiction. □

Kervaire-Milnor defined an abelian group Θ_n for each $n \in \mathbb{Z}^+$ consisting of oriented diffeomorphism classes of oriented differential structures on S^n. The elements in Θ_n are usually called *homotopy spheres*, and the non-zero elements are called *exotic spheres*. (When $n = 4$, Θ_4 consists of h-cobordism classes of smooth structures on S^4.) $\Theta_n = 0$ for $n < 7$, $\Theta_7 = \mathbb{Z}_{28}$. The group operation is connected sum. They showed that Θ_n is finite and defined a cyclic subgroup bP_{n+1} of Θ_n consisting of those spheres which bound a parallelizable manifold. And the quotient group Θ_n/bP_{n+1} is closely related to the cokernel of the stable J-homomorphism $J\colon \pi_n(SO(k)) \to \pi_{n+k}(S^k)$ $(k \gg n)$. Furthermore this quotient group splits back to give a direct sum decomposition of Θ_n whenever n is *not* of the form $2^k - 3$ (Brumfiel 1970). For m odd, $bP_m = 0$ and

$$bP_{4k+2} = \begin{cases} 0, & \text{if } K^{4k+2} \text{ is smoothable,} \\ \mathbb{Z}_2, & \text{if } K^{4k+2} \text{ is not smoothable} \end{cases}$$

where in the second case the generator is ∂M^{4k+2}.

Remark 5.6. M^{4k+2} is parallelizable because of the following fact. Recall that $S^{2k+1} \vee S^{2k+1} \subset M^{4k+2}$ and the normal bundle of each sphere is $\nu_{S^{2k+1}} = TS^{2k+1}$. Therefore $TM|_{S^{2k+1}} = TS^{2k+1} \oplus TS^{2k+1} = \Theta^{4k+2}$.

We now discuss the remaining case bP_{4k}, which is generated by the boundary of a related plumbing construction performed on the "E_8 Dynkin diagram".

Recall the E_8 Dynkin diagram is

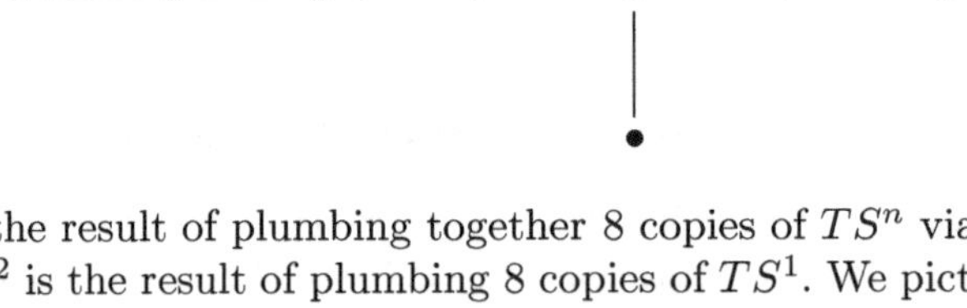

We let $\mathscr{E}^{2n}$ be the result of plumbing together 8 copies of TS^n via this blueprint. For example, $\mathscr{E}^2$ is the result of plumbing 8 copies of TS^1. We picture it in Figure 5.3. Think of making a paper chain link rope with 7 links in a row and off the 5th link attach an extra link. Each link is a cylinder $S^1 \times [0,1]$ (Figure 5.4).

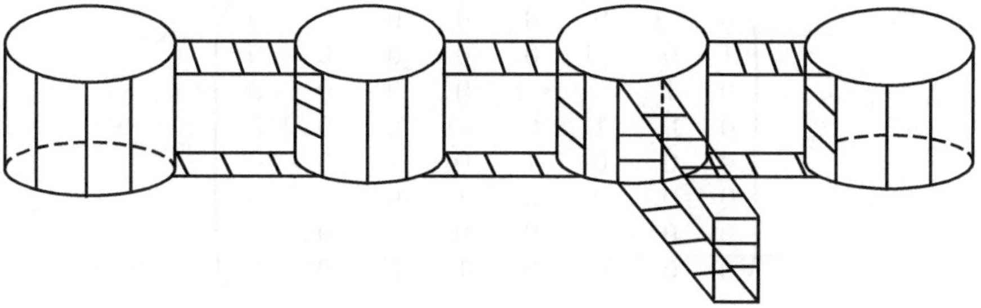

Figure 5.3

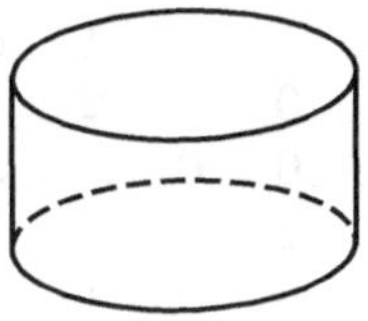

Figure 5.4

Proposition 5.7. *$\partial\mathscr{E}^{2n}$ is homeomorphic to S^{2n-1}, except when $n = 2$.*

Remark 5.8. In particular $\partial\mathscr{E}^2$ is connected (see Figure 5.3). One can try to see this directly. Note that if the extraneous link is missing then the boundary is $S^1 \sqcup S^1$ (top and bottom).

Proof. As in the case of plumbing together 2 copies of TS^n we need only show that $\partial\mathscr{E}^{2n}$ has the same homology groups as S^{2n-1}. (It is simply-connected when $n > 2$.) For $n = 1$ this implies that $\partial\mathscr{E}^2 = S^1$ and for $n > 2$, Smale's Theorem implies that $\partial\mathscr{E}^{2n}$ is homeomorphic to S^{2n-1}.

Now it turns out that $\partial\mathscr{E}^4$ is not simply-connected. In fact $\pi_1(\partial\mathscr{E}^4)$ is the binary icosahedral group A_5^* and $\partial\mathscr{E}^4$ is in fact the "Poincaré fake 3-sphere" S^3/A_5^*. For more details see §6.2.

To complete the proof we must calculate $H_i(\partial\mathscr{E}^{2n})$, which is done, as in the case of plumbing 2 copies of TS^n together, by analyzing the homology exact sequence for $(\mathscr{E}^{2n}, \partial\mathscr{E}^{2n})$. For this the key is to determine the map

$$\tau\colon H_n(\mathscr{E}^{2n}) \to H_n(\mathscr{E}^{2n}, \partial\mathscr{E}^{2n}) \cong H^n(\mathscr{E}^{2n}) \cong \hom(H_n(\mathscr{E}^{2n}), \mathbb{Z}).$$

(Again $\mathscr{E}^{2n} \simeq S^n \vee \cdots \vee S^n$ (8 copies); i.e. a bouquet of eight n-spheres. Hence $\widetilde{H}_i(\mathscr{E}^{2n}) = 0$ for $i \neq n$.)

As before τ can be identified with the linear transformation which is adjoint to the intersection pairing $-\wedge-\colon H_n(\mathscr{E}^{2n}) \times H_n(\mathscr{E}^{2n}) \to \mathbb{Z}$. $\wedge$ is symmetric when n is even and skew symmetric when n is odd. In fact there are only two 8×8-matrices A and B determined by this pairing: A when n is odd and B when n is even.

$$A = \begin{pmatrix} 0 & -1 & 0 & 0 & 0 & 0 & 0 & 0 \\ 1 & 0 & -1 & 0 & 0 & 0 & 0 & 0 \\ 0 & 1 & 0 & -1 & 0 & 0 & 0 & 0 \\ 0 & 0 & 1 & 0 & -1 & 0 & 0 & 0 \\ 0 & 0 & 0 & 1 & 0 & -1 & 0 & -1 \\ 0 & 0 & 0 & 0 & 1 & 0 & -1 & 0 \\ 0 & 0 & 0 & 0 & 0 & 1 & 0 & 0 \\ 0 & 0 & 0 & 0 & 1 & 0 & 0 & 0 \end{pmatrix}$$

$$B = \begin{pmatrix} 2 & 1 & 0 & 0 & 0 & 0 & 0 & 0 \\ 1 & 2 & 1 & 0 & 0 & 0 & 0 & 0 \\ 0 & 1 & 2 & 1 & 0 & 0 & 0 & 0 \\ 0 & 0 & 1 & 2 & 1 & 0 & 0 & 0 \\ 0 & 0 & 0 & 1 & 2 & 1 & 0 & 1 \\ 0 & 0 & 0 & 0 & 1 & 2 & 1 & 0 \\ 0 & 0 & 0 & 0 & 0 & 1 & 2 & 0 \\ 0 & 0 & 0 & 0 & 1 & 0 & 0 & 2 \end{pmatrix}$$

In calculating these matrices we use the core 0-sections of the bundles TS^n, e_i, for the basis elements ordered as follows according to the order of corresponding vertices in the Dynkin diagram (Figure 5.5)

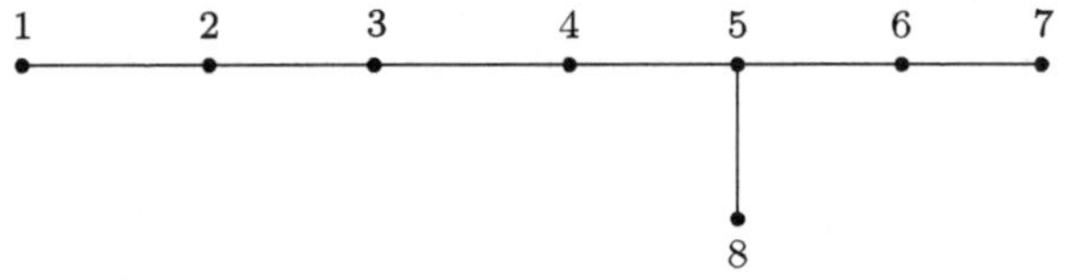

Figure 5.5

Now a standard calculation (by Gauss elimination) shows that $\det A = \det B = 1$. Hence we can repeat the calculation given in the case of plumbing TS^{2k-1} to TS^{2k-1} to show

$$H_i(\partial\mathscr{E}^{2n};\mathbb{Z}) = \begin{cases} \mathbb{Z}, & \text{if } i = 0, 2n-1, \\ 0, & \text{otherwise.} \end{cases}$$

□

Now it is again easy to see that $T\mathscr{E}^{2n}$ is parallelizable. Denote $M^{2n} = \mathscr{E}^{2n}/\partial\mathscr{E}^{2n} = \mathscr{E}^{2n} \cup_{\partial\mathscr{E}^{2n}} \mathbb{D}^{2n}$ where $n > 2$ and $\Sigma^{4n-1} = \partial M^{4n}$ where $n > 1$.

5.2 Surgery

In this section we continue the study of the manifolds $\mathscr{E}^{2n}$ and $\partial\mathscr{E}^{2n}$.

Arguing using ideas due to Kervaire one can show (see the last section of [10] for details) that when n is *odd* there is a different basis for $\mathscr{E}^{2n}$ that can be "killed" by surgery, i.e. four $(2n+1)$-dimensional handles can be attached to $\mathscr{E}^{2n} \times [0,1]$ inside $\mathrm{int}(\mathscr{E}^{2n} \times \{1\})$ yielding a cobordism W^{2n+1} between $\mathscr{E}^{2n}$ and a new smooth manifold $\mathscr{D}^{2n}$ with $\partial\mathscr{D}^{2n} = \partial\mathscr{E}^{2n}$ but such that $\mathscr{D}^{2n}$ is contractible and hence diffeomorphic to $\mathbb{D}^{2n}$. Therefore $\partial\mathscr{E}^{2n} = S^{2n-1}$ when n is odd and M^{2n} has a smooth structure; i.e. nothing new has been produced. The following is a detailed argument for this.

Using Gauss elimination as we saw before one shows that $\det A = \pm 1 = \det B$ and hence the intersection pairing $\wedge$ on $H_n(\mathscr{E}^{2n})$ is *unimodular*; i.e. its adjoint $\mathrm{ad}\wedge\colon H_n(\mathscr{E}^{2n}) \to \hom(H_n(\mathscr{E}^{2n}), \mathbb{Z})$ is an isomorphism.

When n is odd $\wedge$ is skew symmetric and when $n \neq 1, 3, 7$ it is equipped with a $\mathbb{Z}_2$-valued quadratic form $q\colon H_n(\mathscr{E}^{2n}) \to \mathbb{Z}_2$ defined by $q(x) = 0$ or 1 depending on whether the embedded sphere $S^n \subset \mathscr{E}^{2n}$ representing $x \in H_n(\mathscr{E}^{2n}) = \pi_n(\mathscr{E}^{2n})$ has trivial normal bundle ν_{S^n} or not; i.e. $q(x) = 0$ if ν_{S^n} is trivial, $q(x) = 1$ if ν_{S^n} is not trivial. (Any two embeddings representing x are isotopic by Haefliger's extension of the Whitney Embedding Theorem, applicable since $\mathscr{E}^{2n}$ is highly connected.)

Quadratic means that $q(x+y) = q(x) + q(y) + x \wedge y$. Now Arf (1941) studied such $\mathbb{Z}_2$-valued quadratic forms q associated to a $\mathbb{Z}$-valued unimodular skew symmetric bilinear form $b\colon V \times V \to \mathbb{Z}$. First it is easy to see that any such (V, b) has a symplectic basis $e_1, f_1; e_2, f_2; \cdots; e_m, f_m$; i.e. a basis for V such that the corresponding $2m \times 2m$ matrix is blocked diagonal

$$\begin{pmatrix} \begin{pmatrix} 0 & -1 \\ 1 & 0 \end{pmatrix} & 0 & \cdots & \cdots \\ 0 & \begin{pmatrix} 0 & -1 \\ 1 & 0 \end{pmatrix} & 0 & \cdots \\ \vdots & 0 & \ddots & \\ \vdots & \vdots & & \begin{pmatrix} 0 & -1 \\ 1 & 0 \end{pmatrix} \end{pmatrix}$$

i.e. $b(e_i, e_j) = 0$ for all i, j and $b(e_i, f_j) = 0$ if $i \neq j$ and $b(e_i, f_i) = 1$. He then defined a $\mathbb{Z}_2$ invariant $A(q)$ — the Arf invariant — for q associated to b by the formula

$$A(q) = \sum_{i=1}^{m} q(e_i) q(f_i),$$

and showed this is independent of the symplectic basis chosen.

Remark 5.9. Another way of defining $A(q)$ can be found in Browder's book [3, Corollary III.1.9]; namely, since $q(2x) = q(x) + q(x) + b(x,x) = 2q(x) + 0 = 0$, q factors through $\hat{q}\colon V/2V \to \mathbb{Z}_2$. Then define

$$A(q) = \begin{cases} 1, & \text{if the value 1 is assumed more often than the value 0,} \\ 0, & \text{otherwise,} \end{cases}$$

i.e. we always have $|\hat{q}^{-1}(1)| \neq |\hat{q}^{-1}(0)|$ and $A(q) = 1$ if and only if $|\hat{q}^{-1}(1)| > |\hat{q}^{-1}(0)|$.

Also up to isomorphism there are exactly two quadratic forms q associated to b. And if $A(q) = 0$ then there is a symplectic basis such that $q(e_i) = q(f_i) = 0$ for all i. All these things are proven in Browder's book [3].

In the case of $\wedge$ on $H_n(\mathscr{E}^{2n}) = \mathbb{Z}^8$, let $\{a_1, a_2, \cdots, a_8\}$ be the basis from which the matrix A for $\wedge$ was calculated. Note that $q(a_i) = 1$ (Figure 5.6).

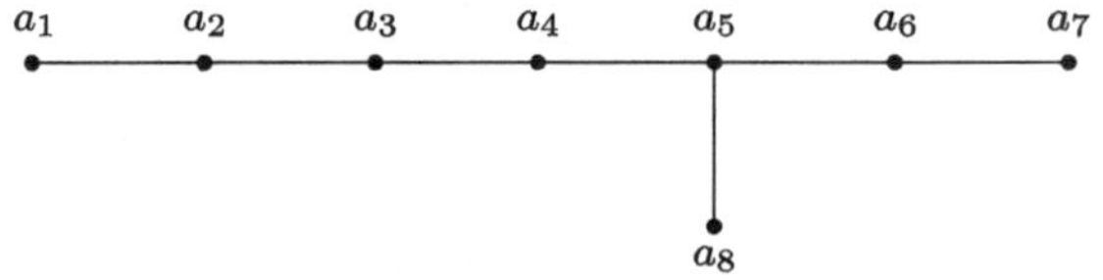

Figure 5.6

Then the following is a symplectic basis for $\wedge$: $e_1 = a_1$, $f_1 = a_2$, $e_2 = a_6$, $f_2 = a_7$, $e_3 = a_1 + a_3$, $f_3 = a_4$, $e_4 = a_1 + a_3 + a_5 + a_7$, $f_4 = a_8$. Hence

$$\begin{aligned} A(q) &= q(e_1)q(f_1) + q(e_2)q(f_2) + q(e_3)q(f_3) + q(e_4)q(f_4) \\ &= 1 + 1 + q(e_3) + q(e_4) \\ &= 0. \end{aligned}$$

Hence we can find a new symplectic basis $\overline{e}_1, \overline{f}_1, \cdots, \overline{e}_4, \overline{f}_4$ such that $q(\overline{e}_i) = 0$; i.e. there exist embeddings $S_i^n \times \mathbb{D}^n \subset \mathscr{E}^{2n}$ representing $\overline{e}_i$ $(i = 1, 2, 3, 4)$. If we attach four $(2n+1)$-handles $\mathbb{D}^{n+1} \times \mathbb{D}^n$ (each of index $n+1$) via $S_i^n \times \mathbb{D}^n$, we get a $(2n+1)$-dimensional cobordism W^{2n+1} with $\partial W = \mathscr{E}^{2n} \cup (\partial \mathscr{E}^{2n} \times [0,1]) \cup \partial^+ W$ such that $\partial^+ W$ is contractible and $\partial(\partial^+ W) = \partial \mathscr{E}^{2n}$ (Figure 5.7). By the h-Cobordism Theorem $\partial^+ W$ is diffeomorphic to $\mathbb{D}^{2n}$ and $\partial \mathscr{E}^{2n}$ is diffeomorphic to S^{2n-1}. So when n is odd, nothing new is obtained.

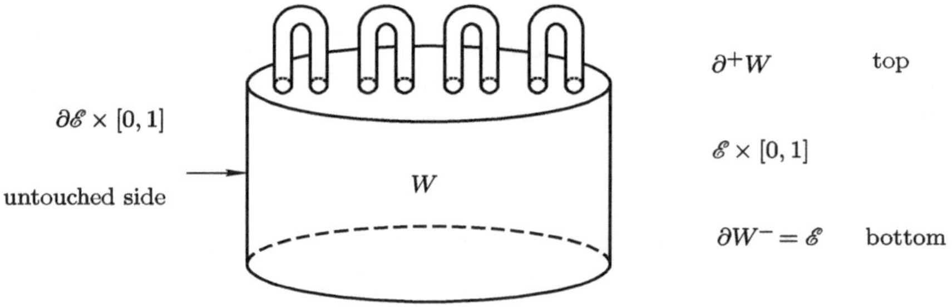

Figure 5.7

Remark 5.10. We can take $\overline{e}_1 = a_1 + a_6$, $\overline{e}_2 = a_2 - a_7$, $\overline{e}_3 = e_3 = a_1 + a_3$, $\overline{e}_4 = e_4 = a_1 + a_3 + a_5 + a_7$.

In dimension 2, we don't have enough room to realize the Whitey Embedding Theorem (so the above arguments do not work verbatim). But it is instructive to note that on Figure 5.3 surgery can be done on $\sigma_1 = a_2$, $\sigma_3 = a_4$, $\sigma_2 = a_7$, $\sigma_4 = a_8$ (part of the noted symplectic basis) since ν is also 0 in this dimension. The result can be visualized to be $\mathbb{D}^2$.

Now we turn to discuss the case when n is even. $E^{2n} = \mathscr{E}^{2n}/\partial\mathscr{E}^{2n}$ usually does *not* have a smooth structure and consequently $\Sigma^{2n-1} = \partial\mathscr{E}^{2n}$ is an "exotic sphere"; i.e. not diffeomorphic to S^{2n-1}. In fact when $n = 4$ it is related to Milnor's original (1956) exotic 7-spheres we discussed in §4.3 (each of them is a connected sum of finitely many copies of Σ^7). To analyze the situation when $n = 2m$ $(m > 1)$ we use the Hirzebruch Index Formula, assuming E^{4m} has a smooth structure to show that this usually leads to a contradiction.

Recall $l_n = a_n p_n +$ sums of products of p_i where $i < n$. And

$$I(E^{4n}) = \langle l_m(E^{4m}), [E^{4m}]\rangle.$$

Also $E^{4m} - \mathbb{D}^{4m} = \text{int}(\mathscr{E}^{4m})$ is parallelizable with the normal smooth structure on $\mathscr{E}^{4m}$. Let α be the smooth structure on E^{4m} posited and $\alpha|_{\mathscr{E}^{4m}}$ be its restriction to $\mathscr{E}^{4m}$. Since the inclusion $\mathscr{E}^{4m} \subset E^{4m}$ induces an isomorphism on rational cohomology in dimension $< 4m$, Novikov's Theorem 4.8 implies that $p_i(E^{4m}) = 0$ in $H^{4i}(E^{4m};\mathbb{Q})$ for all $i < m$. And hence

$$8 = I(E^{4m}) = a_m \langle p_m(E^{4m}), [E^{4m}]\rangle.$$

Recall

$$a_m = \frac{2^{2m}(2^{2m-1}-1)}{(2m)!} B_m \in \mathbb{Q}$$

where B_m is the m-th Bernoulli number. Since $\langle p_m(E^{4m}), [E^{4m}]\rangle \in \mathbb{Z}$, the numerator of a_m *must divide* 8, which is doesn't when $m = 2$. And it can be shown that the numerator of $a_m\langle p_m(E^{4m}), [E^{4m}]\rangle$ never divides 8 when $m \geq 2$. Hence E^{4m} does *not* support a smooth structure for all $m \geq 2$. (See Appendix B of [17].)

Remark 5.11. In the above argument we need to use a result due to Bott (1958) (letters between Bott and Milnor, published as [2]) that $(2m-1)!|\langle p_m(\eta), [S^{4m}]\rangle$ for any bundle η over S^{4m}.

So bP_m is a computable lower bound for Θ_{m-1}. We would like an upper bound; i.e. some way to calculate Θ_{m-1}/bP_m.

For this we "try to define a homomorphism" $\Theta_m \to \pi_{m+k}(S^k)$ $(k \gg m)$, whose kernel is bP_{m+1}. We start by embedding $\Sigma^m \subset S^{m+k}$, where $k > m$, and note that its normal bundle ν_Σ is trivial, since Σ^m is stably parallelizable. (Recall this was shown in the proof of Novikov's Theorem 4.26.) Therefore $\Sigma^m \times \mathbb{D}^k \subset S^{m+k}$. Hence by a natural construction we have used before, there is determined a map $\hat{\Sigma}\colon S^{m+k} \to \mathbb{D}^k/\partial\mathbb{D}^k = S^k$ $(\partial\mathbb{D}^k = \infty)$ defined by

$$\hat{\Sigma}(x) = \begin{cases} \infty, & \text{if } x \notin \Sigma^m \times \text{int}(\mathbb{D}^k), \\ p(x), & \text{if } x \in \Sigma^m \times \text{int}(\mathbb{D}^k), \end{cases}$$

where $p\colon \Sigma^m \times \operatorname{int}(\mathbb{D}^k) \to \operatorname{int}(\mathbb{D}^k)$ is the projection onto the second factor. Note that this definition depends on identifying ν_Σ with $\Sigma^m \times \mathbb{D}^k$. A second trivialization is gotten by using a continuous map $S^m \simeq \Sigma^m \to O(k)$. The result is that $\hat{\Sigma} \in \pi_{m+k}(S^k)$ is only well-defined if we factor $\pi_{m+k}(S^k)$ by the image of the J-homomorphism

$$J\colon \pi_m(O(k)) \to \pi_{m+k}(S^k).$$

So in this way we get a homomorphism

$$\Phi\colon \Theta_m \to \pi_{m+k}(S^k)/\operatorname{im} J, \quad \Sigma \mapsto \hat{\Sigma}.$$

And it is a pleasant exercise to show that $\ker \Phi = bP_{m+1}$. Here are a few details.

Figure 5.8 illustrates why $bP_{m+1} \subseteq \ker \Phi$. In it, W^{m+1} is a parallelizable manifold such that $\partial W = \Sigma^m$. If $k \gg m$, we can embed $W^{m+1} \subset \mathbb{D}^{(m+1)+k}$ such that $\partial\mathbb{D}^{(m+1)+k} \cap W = \partial W = \Sigma^m \subset S^{m+k}$. And the normal bundle ν_W is trivial. Thus $W \times \mathbb{D}^k \subset \mathbb{D}^{(m+1)+k}$ such that

$$W \times \mathbb{D}^k \cap (S^{m+k} = \partial\mathbb{D}^{(m+1)+k}) = \Sigma^m \times \mathbb{D}^k \subset S^{m+k}.$$

Hence the natural geometric construction given above yields a continuous map

$$\hat{W}\colon \mathbb{D}^{(m+1)+k} \to \mathbb{D}^k/\partial\mathbb{D}^k = S^k$$

extending $\hat{\Sigma}\colon \partial\mathbb{D}^{(m+1)+k} \to S^k$.

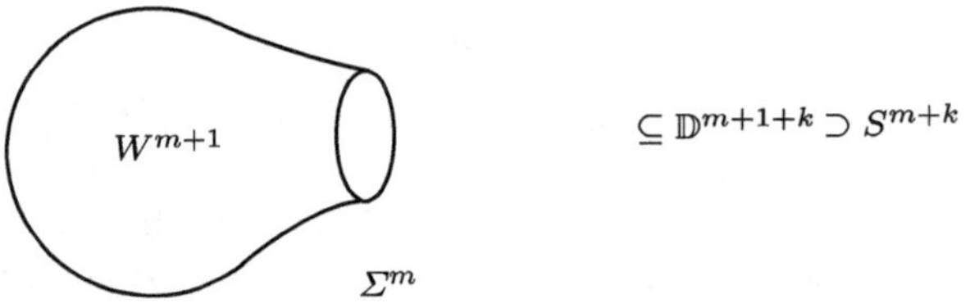

Figure 5.8

On the other hand if $\hat{\Sigma}$ is null homotopic, we proceed to show that $\Sigma^m \in bP_{m+1}$ as follows. Let $f\colon \mathbb{D}^{m+1+k} \to S^k$ extend $\hat{\Sigma}$. Recall $\hat{\Sigma}^{-1}(0) = \Sigma^m$, where $0 \in \mathbb{D}^k/\partial\mathbb{D}^k = S^k$. We can assume that 0 is a regular value of f; hence $W^{m+1} = f^{-1}(0)$ is a parallelizable manifold (note that $\nu_W = f^*(\text{trivial bundle})$) such that $\partial W = \Sigma^m$ (Figure 5.9).

It remains to analyze $\operatorname{im} \Phi \subseteq \pi_{m+k}(S^k)/\operatorname{im} J$. We will show that the quotient is either 0 or C_2, and almost always 0. To do this analysis start with a continuous (smooth) map $f\colon S^{m+k} \to S^k$ which we can assume has $0 \in \mathbb{D}^k/\partial\mathbb{D}^k = S^k$ as a regular value; and let $M^n = f^{-1}(0)$. Note the normal bundle ν_M is trivial. Hence M^n is stably parallelizable. If M^n was simply-connected and had the homology groups of S^n, then M^n would be a smooth manifold homeomorphic to S^n by the theorems of Hurewicz and Smale. And consequently $[f] \in \pi_{m+k}(S^k)$ would be in $\operatorname{im} \Phi$. Now one can always easily do (framed) surgeries on M^n so that the result is

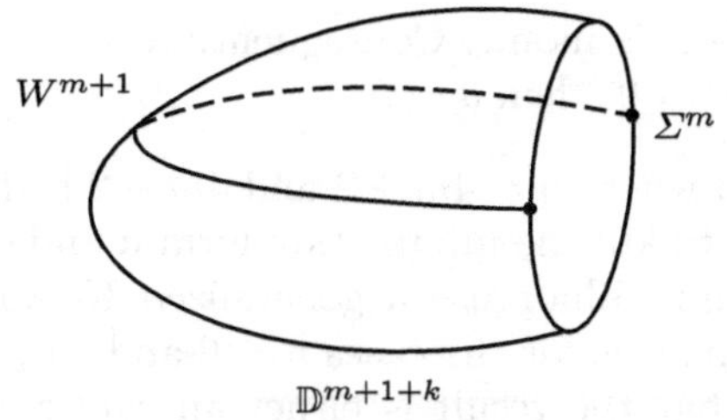

Figure 5.9

simply-connected and $H_i(M^n) = H_i(S^n)$ except possibly in the middle dimension $n/2$ if n is even and $(n-1)/2$, $(n+1)/2$ if n is odd. (And the result doesn't change $[f] \in \pi_{m+k}(S^k)$.) Also when n is odd, by a very hard argument one can always kill by surgery even these middle dimensional groups. Hence when n is odd, $\operatorname{im} \Phi = \pi_{m+k}(S^k)/\operatorname{im} J$.

Also when $n = 4m$, one can kill by surgery the middle dimensional group $H_{2m}(M^{4m}; \mathbb{Z})$ always using the Hirzebruch Index Formula; i.e. $I(M^{4m}) = 0$ since all the Pontryagin classes of M^{4m} vanish. Now using a number theory result the intersection pairing $\wedge$ on $H_{2m}(M^{4m})$ is stably equivalent to one with hyperbolic basis; i.e. a basis e_1, f_1; e_2, f_2; $\cdots$; e_s, f_s represented by a blocked diagonal matrix with 2×2 blocks

$$\begin{pmatrix} 0 & 1 \\ 1 & 0 \end{pmatrix}$$

down the diagonal. By attaching s $(4m+1)$-handles of index $2m+1$, $\mathbb{D}^{2m+1} \times \mathbb{D}^{2m}$, along $e_1, e_2, \cdots, e_s$, the resulting manifold is an exotic sphere Σ^{4m}.

Remark 5.12. Kervaire-Milnor claim it is easier to see that (without stabilizing) any index 0 unimodular symmetric form $\wedge$ has a basis $e_1, \cdots, e_s, f_1, \cdots, f_s$ such that the $2s \times 2s$ matrix representing $\wedge$ in terms of this basis is

$$\begin{pmatrix} 0 & A \\ A & B \end{pmatrix}.$$

This is sufficient to do the surgeries on $e_1, \cdots, e_s$ given the following lemma.

Lemma 5.13. *Let $E \to S^{2m}$ be a $2m$-dimensional vector bundle which is both*

1. *stably trivial; i.e. $E \oplus \Theta^1 = \Theta^{2m+1}$; and*
2. $\chi(E) = 0$.

Then E is trivial; i.e. $E = \Theta^{2m}$.

Proof. Let $s = 2m$ and consider the fiber bundle $O(s) \to O(s+1) \to S^s$ and the induced homotopy exact sequence

$$\pi_s(S^s) \to \pi_{s-1}(O(s)) \to \pi_{s-1}(O(s+1)) \to \pi_{s-1}(S^s) = 0$$

with $\pi_s(S^s) = \mathbb{Z}$ and $\pi_{s-1}(O(s))$ is the set of isomorphism classes of s-plane bundles over S^s. But $\chi(TS^s) = 2$ since s is even. Hence the composite $\mathbb{Z} =$

$\pi_s(S^s) \to \pi_{s-1}(O(s)) \xrightarrow{\chi} \mathbb{Z}$ is monic. Consequently if $a \in \pi_{s-1}(O(s))$ goes to 0 in $\pi_{s-1}(O(s+1))$ and $\chi(a) = 0$, then $a = 0$. □

In the final situation where $n = 4m+2$ and $4m+2$ is different from 6 and 14, we can perform surgery to kill the intersection form if and only if its Arf invariant $A(\wedge) = 0$. (Kervaire and Milnor use a generalized Kervaire invariant, still Arf type, to complete the analysis for the cases $n = 6$ and 14.) In fact one can always do framed surgery so that the result is either an exotic sphere or the Kervaire manifold K^{4m+2}. But the latter is only possible when K^{4m+2} is smoothable, which we now know only happens when $m = 4l+2 = 2, 6, 14, 30, 62$ and possibly 126. Summarizing we have (let $k \gg m \geq 5$)

1. when $m \neq 6, 14, 30, 62$ and 126, $\Theta_m/bP_{m+1} = \pi_{m+k}(S^k)/\operatorname{im} J$;
2. when $m = 6, 14, 30, 62$, Θ_m has index 2 in $\pi_{m+k}(S^k)$;
3. Θ_{126} has index 2 or 1 in $\pi_{126+k}(S^k)$ depending on whether K^{126} is smoothable or not.

Now let us return to the manifold $\mathscr{E}^4$ resulting from plumbing 8 copies of $\mathbb{D}(TS^2)$ according to the Dynkin diagram E_8. Recall its boundary $\partial\mathscr{E}^4 = \Sigma$ has the homology groups of S^3 but is not simply-connected. Freedman, as part of 1982 Fields' Medal work, constructed a compact contractible topological 4-manifold $\mathscr{D}^4$ whose boundary is $\partial\mathscr{D}^4 = \Sigma$. In fact Freedman showed that any "homology 3-sphere" M^3 (meaning M is a closed 3-manifold with $H_*(M) \cong H_*(S^3)$) bounds a compact contractible topological 4-manifold.

Let $E^4 = \mathscr{E}^4 \cup_\Sigma \mathscr{D}^4$. Then

Theorem 5.14 (Freedman). *E^4 is a closed topological 4-manifold which does* not *support a smooth structure.*

Proof. Suppose it does, we will show this assumption leads to a contradiction. First note that $I(E^4) = 8$. This is because the intersection form $\wedge\colon H_2(E^4) \times H_2(E^4) \to \mathbb{Z}$ is the matrix B on Page 80. Note calculating $\wedge$ does not use the given smooth structure. It is a cohomology calculation and is the same as

$$H^2(\mathscr{E}^4/\partial\mathscr{E}^4) \times H^2(\mathscr{E}^4/\partial\mathscr{E}^4) \xrightarrow{\cup} H^4(\mathscr{E}^4/\partial\mathscr{E}^4) = \mathbb{Z}$$

with $H^2(\mathscr{E}^4/\partial\mathscr{E}^4) = \mathbb{Z}^8$. Now recall Rochlin's Theorem (Theorem 4.23) that the index of any almost parallelizable smooth 4-manifold is divisible by 16. Consequently $T(E^4 - *) \not\cong \Theta^4$. Our final step is to contradict this; i.e. we will next show that $T(E^4 - *) \cong \Theta^4$. Since the inclusion $\operatorname{int}(\mathscr{E}^4) \subset E^4 - *$ is a homotopy equivalence (here we pick $* \in \operatorname{int}(\mathscr{D}^4)$), it suffices to show that $T(\operatorname{int}(\mathscr{E}^4), \alpha) = \Theta^4$, where α is the induced differential structure on $\operatorname{int}(\mathscr{E}^4)$ from the posited one on E^4. Note that α may not be diffeomorphic to the *natural* structure on $\operatorname{int}(\mathscr{E}^4)$; i.e. the one resulting from plumbing 8 copies of TS^2 together. Recall the natural structure is clearly parallelizable. However we do have that $\operatorname{int}(\mathscr{E}^4) = \cup_{i=1}^8 U_i$ where each U_i is an open set homeomorphic to TS^2 and whose intersection pattern is given by the *tree*

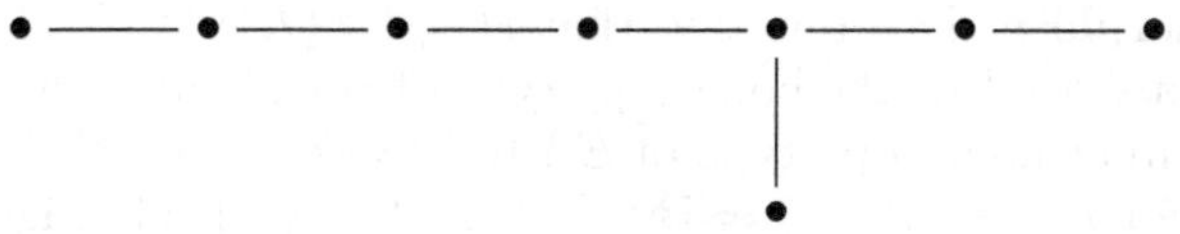

and where $U_i \cap U_j$ is either empty or contractible when $i \neq j$. Therefore it suffices to prove the following

Lemma 5.15. *Let U be a smooth 4-manifold homeomorphic to the total space TS^2, then U is parallelizable.*

Proof. Since TU is stable because U is open, it is enough to show that the 6-dimensional manifold $U \times \mathbb{R}^2$ is parallelizable. Now by the Whitney Embedding Theorem we have a smooth embedding $\sigma\colon S^2 \subset U \times \mathbb{R}^2$ so that σ is a homotopy equivalence. Hence by Smale's h-Cobordism Theorem and Browder's form of the splitting theorem $\nu_\sigma = U \times \mathbb{R}^2$; i.e. the total space of ν_σ is diffeomorphic to $U \times \mathbb{R}^2$. And recall that $T(\nu_\sigma)|_{S^2} = TS^2 \oplus \nu_\sigma$, which is stably isomorphic to ν_σ. In this way, we have reduced the lemma to showing that $\nu_\sigma = \Theta^4$. Recall the Thom complex of a vector bundle E (denoted here by $\mathscr{T}(E)$) is the one point compactification of its total space. Therefore

$$\mathscr{T}(\nu_\sigma) = U \times \mathbb{R}^2 \cup \infty = (TS^2 \times \mathbb{R}^2) \cup \infty = \mathscr{T}(TS^2 \oplus \Theta^2) = \mathscr{T}(\Theta^4) \simeq S^6 \vee S^4;$$

i.e. $\mathscr{T}(\nu_\sigma)$ is homotopy equivalent to $S^6 \vee S^4$. Now recall that $\widetilde{KO}(S^2) = \mathbb{Z}_2$; i.e. there are only two stable bundles over S^2, namely Θ^4 and $\eta \oplus \Theta^2$, where η is the canonical complex line bundle over $\mathbb{C}\mathrm{P}^1 = S^2$. Now $\mathscr{T}(\eta) = \mathbb{C}\mathrm{P}^2$ and consequently $\mathscr{T}(\eta \oplus \Theta^2) = \Sigma^2\mathbb{C}\mathrm{P}^2$. And we see that $\Sigma^2\mathbb{C}\mathrm{P}^2$ is not homotopy equivalent to $S^6 \vee S^4$, by considering the Steenrod square $Sq^2\colon H^4(-;\mathbb{Z}_2) \to H^6(-;\mathbb{Z}_2)$. Recall that the Steenrod squares are "stable cohomology operations"; i.e. they commute with the suspension isomorphisms

$$\begin{array}{ccc} H^i(X) & \xrightarrow{\Sigma} & H^{i+1}(\Sigma X) \\ \downarrow{\scriptstyle Sq^n} & & \downarrow{\scriptstyle Sq^n} \\ H^{i+n}(X) & \xrightarrow{\Sigma} & H^{i+n+1}(\Sigma X) \end{array}$$

Now $Sq^2(\Sigma^2 a) = \Sigma^2 Sq^2(a) = \Sigma^2 a^2$ which generates $H^6(\Sigma^2\mathbb{C}\mathrm{P}^2;\mathbb{Z}_2)$, here $a \in H^2(\mathbb{C}\mathrm{P}^2;\mathbb{Z}_2)$ is a generator and a^2 generates $H^4(\mathbb{C}\mathrm{P}^2;\mathbb{Z}_2)$ and

$$Sq^2\colon H^4(S^6 \vee S^4;\mathbb{Z}_2) \to H^6(S^6 \vee S^4;\mathbb{Z}_2)$$

is clearly zero. Therefore $\mathscr{T}(\nu_\sigma)$ is not homotopy equivalent to $\mathscr{T}(\eta \oplus \Theta^2)$. Consequently $\nu_\sigma \cong \Theta^4$. □

This finishes the proof of Theorem 5.14. □

Remark 5.16. If E is an n-dimensional vector bundle over a compact space B, $\mathscr{T}(E)$ is the one point compactification of E; i.e. $\mathscr{T}(E) = E \cup \infty = \mathbb{D}(E)/S(E)$.

The Thom Isomorphism Theorem says that $\widetilde{H}_{i+n}(\mathscr{T}(E)) \cong H_i(B)$ regardless of E (when E is orientable). But the homotopy type of $\mathscr{T}(E)$ can vary with E. (It is an invariant of the fiber homotopy type of E.) If $E = \Theta^n$, then $\mathscr{T}(\Theta^n) = S^n \vee \Sigma^n B$. Here is a proof for $n = 1$: $\mathscr{T}(\Theta^1) = B \times [-1,1]/B \times \{-1,1\}$ (Figure 5.10).

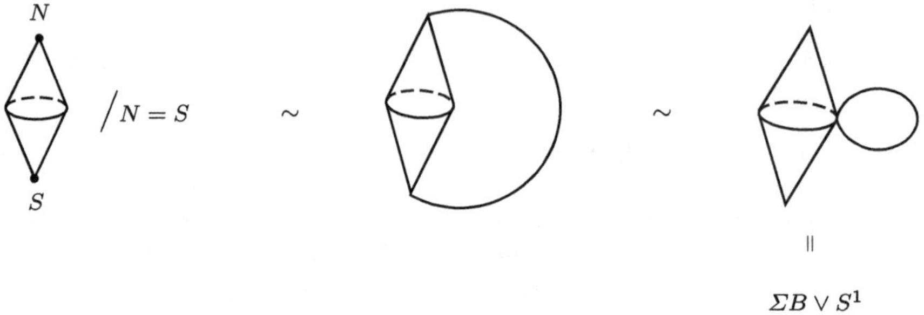

Figure 5.10

Corollary 5.17. *E^4 can not be triangulated.*

Remark 5.18. Freedman "proved" this result in his fundamental 1982 JDG paper [5] but *modulo the Poincaré Conjecture being true*! Since the Poincaré Conjecture has now been established, Freedman's argument is complete. About 5 years after Freedman's paper, Casson gave an independent proof that E^4 is a non-triangulable 4-manifold. So the credit for the first proof of Corollary 5.17 is due to Casson.

Proof. Here is a sketch of Freedman's argument. Again we proceed via proof by contradiction and assume that E^4 has a triangulation K. If the link of each vertex of K is homeomorphic to S^3, then this is a so-called "PL triangulation" of E^4. (See §6.3 for the definition. This uses the fundamental work of Moise on existence and uniqueness of triangulations of 3-manifolds.) But for manifolds M with a PL-structure Hirsch and Mazur (1964 notes, 1974 published [8]) constructed an obstruction theory with obstructions in $H^i(M; \Theta_{i-1})$ for M to have a smooth structure. Since $\Theta_i = 0$ for $i < 7$, it is a consequence that every PL-manifold of dim ≤ 7 has a smooth structure. (*Remark:* The Hirsch-Mazur result uses group Γ_i which are now known to be the same as Θ_i; c.f. [17, pp. 250–251].) Since E^4 has no smooth structure, we conclude that E^4 has no PL-structure and consequently there exists a vertex $v \in K$ such that $Lk(v)$ is *not* homeomorphic to S^3. But $Lk(v)$ is a triangulated homology 3-manifold; i.e. for each $x \in Lk(v)$

$$H_i(Lk(v), Lk(v) - x) \cong H_i(\mathbb{D}^3, \mathbb{D}^3 - 0).$$

It is known that homology 3-manifolds are 3-manifolds. Furthermore, by an argument given in the proof of Assertion 6.22, since E^4 is a manifold, $\pi_1(Lk(v)) = 1$. Therefore by the Poincaré Conjecture $Lk(v)$ is homeomorphic to S^3. This contradiction establishes the corollary. □

6

Hauptvermutung

6.1 The Fundamental Theorem of algebraic K-theory

The splitting Theorem 4.27 and fibering Theorem 4.29 depend on the following basic algebraic result.

Theorem 6.1. *The Whitehead group of a free abelian group of rank n is zero; i.e.*

$$Wh(C^n) = 0$$

where $C^n = C \oplus C \oplus \cdots \oplus C$ *is the sum of* n *copies of the infinite cyclic group* C.

Remark 6.2. For $n = 1$, this was due to Higman (1940, Ph. D. thesis); for $n > 1$ it is due to Bass-Heller-Swan (1964) and is a consequence of the following result.

Fundamental Theorem of algebraic K-theory (Bass-Heller-Swan). *Let* R *be a ring with* 1, *then*

$$K_1(R[C]) = K_1(R) \oplus K_0(R) \oplus Nil(R) \oplus Nil(R).$$

In the first part of this section we will sketch some key points in the proof of this theorem.

There are two ways of defining $Nil(R)$. We start with its matrix description. Let $\mathfrak{Nil}(R)$ denote the free abelian group on the set of all square nilpotent matrices N (i.e. $N^s = 0$ for some $s \in \mathbb{Z}^+$) with entries from R. Then $Nil(R)$ is a quotient group of $\mathfrak{Nil}(R)$ by the following relations where $[N]$ denotes the element in $Nil(R)$ represented by the $n \times n$ nilpotent matrix N:

1. $[N] = [ANA^{-1}]$ for every $A \in GL_n(R)$;
2. $[0] = 0$;
3. if both A and C are nilpotent, then

$$\begin{bmatrix} A & B \\ 0 & C \end{bmatrix} = [A] + [C].$$

(Note that if A and C are nilpotent, then

$$\begin{pmatrix} A & B \\ 0 & C \end{pmatrix}$$

is also nilpotent.)

One can show the following 3 facts.

Fact 1. Every element in $Nil(R)$ is represented by a nilpotent matrix.

Fact 2. Stability

$$[N] = \begin{bmatrix} N & 0 \\ 0 & 0 \end{bmatrix}.$$

Fact 3. $[N] = 0$ if and only if N is stably conjugate to a strictly upper triangular matrix (i.e. one with 0's on the diagonal).

Stallings showed that $K_0(R)$ has a similar description in terms of idempotent matrices P (i.e. square matrices with R-entries such that $P^2 = P$); namely $K_0(R)$ is the quotient group of the free abelian group $\mathfrak{Idem}(R)$ with basis $\{P \mid P \text{ idempotent}\}$ via the relations:

1. if P is an $n \times n$ idempotent matrix, then $[APA^{-1}] = [P]$ for $A \in GL_n(R)$;
2. $[0] = 0$;
3. if

$$\begin{pmatrix} P & S \\ 0 & Q \end{pmatrix}$$

is idempotent, then

$$\begin{bmatrix} P & S \\ 0 & Q \end{bmatrix} = [P] + [Q].$$

And one has the following facts:

Fact 1.

$$[P] = \begin{bmatrix} P & 0 \\ 0 & 0 \end{bmatrix}.$$

Fact 2. $[P] = 0$ in $\widetilde{K}_0(R)$ if and only if P is stably conjugate to

$$\begin{bmatrix} I & 0 \\ 0 & 0 \end{bmatrix}$$

where the size of I depends on P.

With these descriptions, we may define a homomorphism of abelian groups

$$K_1(R) \oplus K_0(R) \oplus Nil(R) \oplus Nil(R) \to K_1(R[C])$$

as follows:

Let t be a fixed generator for C; hence t^{-1} is the other generator. Note that if N is an $n \times n$ nilpotent matrix over R then $I + tN$ is an invertible matrix over $R[C]$; namely its inverse is

$$\sum_{i=0}^{\infty}(-1)^i t^i N^i = I - tN + t^2N^2 - t^3N^3 + \cdots$$

where this is a *finite sum* since $N^s = 0$ for s large enough. This is the standard formula for

$$\frac{1}{1-x} = 1 + x + x^2 + \cdots .$$

It uses crucially that t commutes with every element in $R[C]$.

Similarly $I + t^{-1}N$ is also invertible. This gives two group homomorphisms

$$Nil(R) \to K_1(R[C])$$

by sending N to $I + tN$ or to $I + t^{-1}N$, which yield the two summands $Nil(R) \oplus Nil(R)$ in the theorem.

Now if P is an $n \times n$ idempotent matrix, note that so is $I - P$ since

$$(I-P)^2 = (I-P)(I-P) = I - P - P + P^2 = I - P.$$

Hence $tP + (I - P)$ is invertible with inverse $t^{-1}P + (I - P)$, which is seen via the following simple calculation

$$\begin{aligned}
&(tP + (I-P))(t^{-1}P + (I-P)) \\
&= (tP)(t^{-1}P) + (I-P)(I-P) + (I-P)t^{-1}P + tP(I-P) \\
&= P^2 + I - P + 0 + 0 \\
&= I.
\end{aligned}$$

Hence $P \mapsto tP + (I - P)$ gives a homomorphism of $K_0(R) \to K_1(R[C])$ which gives the summand of $K_0(R)$ in the theorem.

Finally the summand $K_1(R)$ in $K_1(R[C])$ comes functorially by regarding $GL_n(R) \subset GL_n(R[C])$.

Remark 6.3. We may show that these homomorphisms are well-defined by verifying that the relations are preserved. Note that in $K_1(R)$ we have the relation

$$\begin{bmatrix} A & B \\ 0 & C \end{bmatrix} = [A] + [C],$$

since

$$\begin{aligned}
\begin{pmatrix} A & B \\ 0 & C \end{pmatrix} &= \begin{pmatrix} A & 0 \\ 0 & C \end{pmatrix}\begin{pmatrix} I & A^{-1}B \\ 0 & I \end{pmatrix} \\
&= \begin{pmatrix} A & 0 \\ 0 & I \end{pmatrix}\begin{pmatrix} I & 0 \\ 0 & C \end{pmatrix} \cdot \text{products of elementary matrices.}
\end{aligned}$$

Bass Projection

In verifying the Bass-Heller-Swan formula we still need to show that $K_0(R)$, $Nil^{\pm}(R)$ inject into $K_1(R[C]) = K_1(R[t,t^{-1}])$. For this, we need a map

$$\beta\colon K_1(R[t,t^{-1}]) \to K_0(R) \oplus Nil(R)$$

which we call the "Bass projection" (the Bass projection is also topologically interesting), whose description is given in terms of a "new definition" of $K_0(R) \oplus Nil(R)$ as the Grothendieck construction applied to the category $\mathscr{C}(R)$ of nilpotent endomorphisms of finitely generated projective R-modules; i.e. a pair (P,φ) is an object of $\mathscr{C}(R)$ if P is a finitely generated projective R-module, φ is an endomorphism of P, and $\varphi^n = 0$ for some $n \in \mathbb{Z}^+$. A map $f\colon (P,\varphi) \to (Q,\psi)$ is an R-module homomorphism $f\colon P \to Q$ such that $\psi \circ f = f \circ \varphi$; i.e. the diagram below commutes

$$\begin{array}{ccc} P & \xrightarrow{f} & Q \\ \downarrow{\scriptstyle\varphi} & & \downarrow{\scriptstyle\psi} \\ P & \xrightarrow{f} & Q. \end{array}$$

Short exact sequences in $\mathscr{C}(R)$ make sense as follows:

$$0 \to (P,\varphi) \xrightarrow{f} (Q,\psi) \xrightarrow{g} (S,\eta) \to 0 \tag{6.1}$$

is short exact in $\mathscr{C}(R)$ if $0 \to P \xrightarrow{f} Q \xrightarrow{g} S \to 0$ is short exact as a sequence of R-modules.

We define an abelian group $C(R)$ to be the quotient of the free abelian group with "basis the objects" in $\mathscr{C}(R)$ where we factor out the following relation

$$[(P,\varphi)] + [(S,\eta)] = [(Q,\psi)]$$

for any short exact sequence (6.1).

Caveat 6.4. *Quillen develops his higher K-theory by generalizing this categorical setup. But a slight logic concern is always present; i.e. the collection of objects is usually too big to be a set. So one usually uses "small" categories; i.e. those where this collection is a set.*

Forgetting the nilpotent endomorphisms clearly defines a homomorphism $C(R) \to K_0(R)$, $[(P,\varphi)] \mapsto [P]$. And a splitting homomorphism back is induced by $[P] \mapsto [(P,0)]$.

On the other hand, if N is an $m \times m$ nilpotent matrix, then $(R^m, N) \in \mathrm{obj}\mathscr{C}(R)$. But since we required $[0] = 0$ in $Nil(R)$, we define a homomorphism $Nil(R) \to C(R)$ by

$$[N] \mapsto [(R^m, N)] - [(R^m, 0)].$$

To define a splitting backwards, given $(P, v) \in \mathrm{obj}\mathscr{C}(R)$, choose a finitely generated projective module Q such that $P \oplus Q = R^m$ and let N be the nilpotent matrix representing $v\oplus 0\colon P\oplus Q \to P\oplus Q$ with respect to the basis of R^m. Then $[(P,v)] \mapsto [N]$ gives the desired splitting. In this way we see that $C(R) = K_0(R) \oplus Nil(R)$.

We now proceed to define

$$\beta\colon K_1(R[t,t^{-1}]) \twoheadrightarrow C(R)$$

in such a way that $K_0(R) \oplus Nil^-(R)$ maps isomorphically to $C(R)$. Let F denote the free R-module R^m of rank m. Then

$$\overline{F} = \sum_{i=-\infty}^{\infty} Ft^i,\ F^+ = \sum_{i=0}^{\infty} Ft^i,\ F^- = \sum_{i=-\infty}^{0} Ft^i$$

are the corresponding free $R[t,t^{-1}]$-, $R[t]$- and $R[t^{-1}]$-modules of rank m. Let $A \in GL_m(R[t,t^{-1}])$, then for $n \in \mathbb{Z}^+$ big enough $t^nA \in M_m(R[t])$, although probably *not* in $GL_m(R[t])$. In any event the quotient $P = F^+/t^nA(F^+)$ is an $R[t]$-module and hence P is an R-module and the multiplication by t is an R-module endomorphism v of P. We claim that $(P,v) \in \mathrm{obj}\mathscr{C}(R)$. To show this, let $B = A^{-1} \in GL_m(R[t,t^{-1}])$ and apply the above construction to it; i.e. let $p \in \mathbb{Z}^+$ be such that $t^pB \in M_m(R[t])$. Now abbreviate $t^nA\colon F^+ \to F^+$ by f and $t^pB\colon F^+ \to F^+$ by g. Then $g\circ f = f\circ g = t^{n+p}$. Since $g(F^+) \subset F^+$, $t^{n+p}(F^+) = fg(F^+) \subset f(F^+) \subset F^+$, and we get the following short exact sequence of $R[t]$-modules

$$0 \to \frac{F^+}{g(F^+)} \cong \frac{f(F^+)}{fg(F^+)} \to \frac{F^+}{t^{p+n}F^+} \to \frac{F^+}{f(F^+)} = P \to 0. \tag{6.2}$$

From this, we see that P is a finitely generated R-module since $\mathscr{P} = F^+/t^{p+n}F^+ \cong R^{m(n+p)}$ as an R-module. Also note that $t\colon \mathscr{P} \to \mathscr{P}$ is represented by an upper triangular matrix and hence the pair $(\mathscr{P},t)$ is equivalent to $(\mathscr{P},0)$ in $C(R)$. Also this implies that $v\colon P \to P$ is nilpotent, since it is the image of $(\mathscr{P},t)$ under the surjection in the sequence (6.2), $(\mathscr{P},t) \mapsto (P,v)$. It remains to show that P is projective. For this consider $f(F^+) \subset F^+ \subset \overline{F}$; taking quotient we get

$$0 \to \frac{F^+}{f(F^+)} \to \frac{\overline{F}}{f(F^+)} \to \frac{\overline{F}}{F^+} \to 0$$

which is exact as R-modules. Since $\overline{F}/f(F^+) \cong \overline{F}/F^+$, which is clearly free (but infinitely generated), P is a projective R-module. Hence we define

$$\beta([A]) = [(P,v)] - [(R^{nm},0)].$$

The last term accounts for t^n.

To show that β is well-defined and additive, i.e. $\beta([AC]) = \beta([A]) + \beta([C])$, argue as in the construction of the exact sequence (6.2). To show that the composition

$$Nil(R) \to K_1(R[t^{-1}]) \to K_1(R[t,t^{-1}]) \xrightarrow{\beta} C(R) \to Nil(R)$$

is an isomorphism, note

$$N \mapsto I + t^{-1}N \mapsto t(I + t^{-1}N) = tI + N \mapsto \left(\frac{F^+}{f(F^+)}, v\right) \cong (F, -N) \mapsto -N.$$

And to see that the composition

$$K_0(R) \to K_1(R[t, t^{-1}]) \xrightarrow{\beta} C(R) \to K_0(R)$$

is an isomorphism, consider

$$A \mapsto tA + (I - A) \mapsto \beta(tA + (I - A))$$

where $A^2 = A$. Now $F = P \oplus Q$ where $P = \operatorname{im} A$ and $Q = \operatorname{im}(I - A)$. Let

$$P^+ = \sum_{i=0}^{\infty} t^i P, \quad Q^+ = \sum_{i=0}^{\infty} t^i Q,$$

then

$$F^+ = \sum_{i=0}^{\infty} t^i P \oplus \sum_{i=0}^{\infty} t^i Q = P^+ \oplus Q^+$$

and $f(F^+) = (tA + (I - A))(P^+ \oplus Q^+) = tP^+ \oplus Q^+$, hence

$$\frac{F^+}{f(F^+)} = P^+/tP^+ = P.$$

Therefore $\beta(tA + (I - A)) = (P, 0)$, hence under $K_0(R) \to K_0(R)$ the idempotent A goes to its image P which is Stallings' way of identifying the idempotent with the projective module description of $K_0(R)$.

We will come back to using the Bass projection in proving Theorem 6.48 (via explaining the fibering and splitting Theorems 4.29 and 4.27), giving an h-cobordism W which is not topologically a product and the resulting $X = W/(\partial^+ W = +\infty, \partial^- W = -\infty)$ a topological space with two inequivalent triangulations contradicting the Hauptvermutung. The original counter-example to the Hauptvermutung is due to Milnor (1961) [16]. The Hsiang-Farrell example is not new but a conceptually easier explanation of the one due to Stallings (1965).

Bass-Heller-Swan also proved that $K_1(R[t]) = K_1(R) \oplus Nil(R)$. The proof of this is similar as above. First of all, $K_1(R)$ embeds functorially into $K_1(R[t])$ induced by two ring homomorphisms

$$R \xrightarrow{i} R[t] \xrightarrow{\varepsilon} R,$$

where i is the inclusion and ε sends t to 0. And there is a group homomorphism

$$\varphi\colon Nil(R) \to K_1(R[t])$$

as before, induced by $N \mapsto I + tN$ for nilpotent N and $(I + tN)^{-1} = I - tN + t^2N^2 - t^3N^3 + \cdots$.

Let $\overline{Nil}(R) = \operatorname{im}\varphi$. To show that $\varphi\colon Nil(R) \to \overline{Nil}(R)$ is monic we also use the Bass projection $\beta\colon K_1(R[t,t^{-1}]) \to Nil(R)$ composed with the functorially induced homomorphism $\alpha_*\colon K_1(R[t]) \to K_1(R[t,t^{-1}])$, where $\alpha\colon R[t] \to R[t,t^{-1}]$ is the ring homomorphism determined by $\alpha(t) = -t^{-1}$. Then $\beta \circ \alpha_*$ is a left inverse to φ; i.e. $(\beta \circ \alpha_*) \circ \varphi = \mathrm{id}$.

Clearly $K_1(R) \cap \overline{Nil}(R) = 0$ since $\varepsilon_*(\overline{Nil}(R)) = 0$ and $\varepsilon_*|_{K_1(R)}$ is 1-1. So it remains to show that $K_1(R) + \overline{Nil}(R) = K_1(R[t])$; i.e. $\ker\varepsilon_* = \overline{Nil}(R)$. For this we use "Higman's trick" (from his 1940 Ph. D. Thesis). Clearly any $x \in \ker\varepsilon_*$ can be written as

$$x = [I + A_1 t + A_2 t^2 + \cdots + A_m t^m].$$

We proceed to show after stabilization that if $m > 1$, then x can also be represented by a polynomial of smaller degree $m-1$ as follows (Higman's trick)

$$\begin{aligned} x &= \left[\begin{pmatrix} I + A_1 t + \cdots + A_m t^m & 0 \\ 0 & I \end{pmatrix}\right] \\ &= \left[\begin{pmatrix} I + A_1 t + \cdots + A_{m-1} t^{m-1} + A_m t^m & tI \\ 0 & I \end{pmatrix}\right] \\ &= \left[\begin{pmatrix} I + A_1 t + \cdots + A_{m-1} t^{m-1} & tI \\ -A_m t^{m-1} & I \end{pmatrix}\right]. \end{aligned}$$

Iterating this we get that $x = [I + At]$. We now need only to show that A is nilpotent. For this, let $B_0 + B_1 t + B_2 t^2 + \cdots + B_n t^n = (I + At)^{-1}$, thus

$$I = (I + At)(B_0 + B_1 t + B_2 t^2 + \cdots + B_n t^n),$$

and we get $B_0 + (B_1 + AB_0)t + \cdots + (B_n + AB_{n-1})t^n + AB_n t^{n+1} = I$. Thus

$$B_0 = I, B_1 + AB_0 = 0, \cdots, B_n + AB_{n-1} = 0, AB_n = 0.$$

This implies $B_1 = -A$, $B_2 = A^2$, $\cdots$, $B_n = (-1)^n A^n$ and $A^{n+1} = 0$.

The case of the Fundamental Theorem of algebraic K-theory of particular relevance to topology is where R is the integral group ring $\mathbb{Z}\Gamma$ of a group Γ. An elementary observation is that then $R[C] = \mathbb{Z}[\Gamma \times C]$. And the formula becomes

$$K_1(\mathbb{Z}[\Gamma \times C]) = K_1(\mathbb{Z}\Gamma) \oplus K_0(\mathbb{Z}\Gamma) \oplus Nil(\mathbb{Z}\Gamma) \oplus Nil(\mathbb{Z}\Gamma).$$

And topologists are particularly interested *not* in $K_1(\mathbb{Z}\Gamma)$ but rather in its quotient $Wh(\Gamma) = K_1(\mathbb{Z}\Gamma)/\pm\Gamma$. Taking into account this quotient the above formula becomes

$$Wh(\Gamma \times C) = Wh(\Gamma) \oplus \widetilde{K}_0(\mathbb{Z}\Gamma) \oplus Nil(\mathbb{Z}\Gamma) \oplus Nil(\mathbb{Z}\Gamma). \tag{6.3}$$

Recall the definition of $\widetilde{K}_0$ which can be found in the last paragraph of Chapter 3.

Using this formula, one can inductively calculate $Wh(C^n)$, induction on n, starting with $Wh(C^0) = Wh(1) = 0$, provided one can calculate $\widetilde{K}_0(\mathbb{Z}C^n)$ and $Nil(\mathbb{Z}C^n)$. Fortunately the following two results give sufficient information on these to implement this strategy.

Definition 6.5. *A regular ring is a (left) Noetherian ring of finite global dimension.*

Recall R is *Noetherian* if every submodule of a finitely generated R-module is finitely generated. And R has *finite global dimension* if every R-module M has a projective resolution of some finite length n, where n is independent of M; i.e. there exists an acyclic chain complex of R-modules of length n

$$0 \to P_n \to P_{n-1} \to \cdots \to P_0 \to M \to 0$$

where P_i is projective. The minimum of such n is called the global dimension of R.

Example 6.6. The Hilbert Basis Theorem implies that $\mathbb{Z}C^n$ is Noetherian, and the Hilbert Syzygy Theorem implies that the global dimension of $\mathbb{Z}C^n$ is $n+1$. Hence $\mathbb{Z}C^n$ is a regular ring.

Theorem 6.7 (Serre). *If R is a regular ring, then $K_0(R[C]) = K_0(R)$.*

Theorem 6.8 (Bass). *If R is a regular ring, then $Nil(R) = 0$.*

Remark 6.9. Both these theorems can be found in Bass' book [1] around pp. 600. (Rosenberg's book [18] is also a reference.) Unfortunately Bass' book has no index. Also Bass misattributes Serre's theorem to Grothendieck. Serre's theorem is the result of Serre's effort in proving "Serre Conjecture" that every finitely generated projective $\mathbb{Z}[x_1, \ldots, x_n]$-module is free which was affirmatively solved by Quillen and Suslin independently.

Since $\mathbb{Z}C^n$ is regular, Serre's theorem implies that

$$K_0(\mathbb{Z}[C^n]) = K_0(\mathbb{Z}[C^{n-1}][C]) = K_0(\mathbb{Z}[C^{n-1}]) = \cdots = K_0(\mathbb{Z}[C]) = K_0(\mathbb{Z}) = C,$$

the last identity is because $\mathbb{Z}$ is a principal ideal domain. Therefore every finite generated projective $\mathbb{Z}C^n$-module is *stably* free, hence $\widetilde{K}_0(\mathbb{Z}C^n) = 0$.

Therefore Eq. (6.3) applied to $\Gamma = C^{n-1}$ yields

$$\begin{aligned} Wh(C^n) &= Wh(C^{n-1}) \oplus \widetilde{K}_0(\mathbb{Z}C^{n-1}) \oplus Nil(\mathbb{Z}C^{n-1}) \oplus Nil(\mathbb{Z}C^{n-1}) \\ &= Wh(C^{n-1}) \oplus \widetilde{K}_0(\mathbb{Z}C^{n-1}) \text{ (by Bass)} \\ &= Wh(C^{n-1}) \text{ (by Serre).} \end{aligned}$$

And hence by induction on n

$$Wh(C^n) = Wh(C^0) = Wh(1) = 1.$$

(The last identity is by Gauss elimination.)

We complete this section with some comments on $Nil(R)$ and $\widetilde{K}_0(R)$.

Remark 6.10. It's known that $\widetilde{K}_0(\mathbb{Z}C_{23}) \neq 0$. Also Bass-Murthy showed that $Nil(\mathbb{Z}\Gamma)$ is *not* finitely generated for $\Gamma = C \times A$, where A is finite abelian and the cardinality of A is divisible by a square (e.g. $A = C_4$ or $C_2 \times C_2$). Consequently, by the Fundamental Theorem, $Wh(G)$ is *not* finitely generated when G is a finitely generated abelian group whose torsion subgroup has order divisible by a square and rank$(G) \geq 2$; i.e. $G = C^k \times A$ where A is finite abelian, $|A|$ is divisible by a square and $k \geq 2$.

Theorem 6.11 (Farrell, Proc. AMS 1977). *If $Nil(R) \neq 0$, then it is not finitely generated.*

Here we sketch the proof of Theorem 6.11. We proceed as follows. Let $\sigma \colon R[t^n] \subset R[t]$ be the inclusion. Define a transfer map $\sigma^* \colon K_1(R[t]) \to K_1(R[t^n])$ by

$$\sigma^*(A_0 + A_1 t + \cdots + A_{n-1}t^{n-1}) = \begin{pmatrix} A_0 & A_{n-1}t^n & A_{n-2}t^n & \cdots & A_1 t^n \\ A_1 & A_0 & A_{n-1}t^n & \cdots & A_2 t^n \\ \vdots & \vdots & \vdots & & \vdots \\ A_{n-1} & A_{n-2} & A_{n-3} & \cdots & A_0 \end{pmatrix}$$

where the entries in the matrices A_i (for $i = 0, 1, 2, \cdots, n-1$) are from $R[t^n]$.

Step1. $\sigma^*\sigma_* \colon K_1(R[t^n]) \to K_1(R[t^n])$ is

$$A_0 \mapsto \begin{pmatrix} A_0 & 0 & 0 \\ & \ddots & \\ 0 & 0 & A_0 \end{pmatrix},$$

consequently $\sigma^*\sigma_*$ is multiplication by n.

Step2. $\sigma_*(Nil_n(R)) \subseteq Nil(R)$, where $Nil_n(R)$ is the image of $Nil(R)$ in $K_1(R[t^n])$ given by $N \mapsto I + t^n N$.

Proof. $I + t^n N \in GL(R[t])$ is in $\ker \varepsilon_*$. □

Step3. From the definitions we have that

$$\begin{aligned} \sigma^*(I + tN) &= \begin{bmatrix} I & 0 & 0 & \cdots & 0 & t^n N \\ N & I & 0 & \cdots & 0 & 0 \\ 0 & N & I & \cdots & 0 & 0 \\ \vdots & \vdots & \vdots & & \vdots & \vdots \\ 0 & 0 & 0 & \cdots & I & 0 \\ 0 & 0 & 0 & \cdots & N & I \end{bmatrix} \\ &= I + M + A \end{aligned}$$

where

$$A = \begin{bmatrix} 0 & \cdots & 0 & t^n N \\ 0 & \cdots & 0 & 0 \\ \vdots & & \vdots & \vdots \\ 0 & \cdots & 0 & 0 \end{bmatrix},$$

$$M = \begin{bmatrix} 0 & 0 & 0 & \cdots & 0 & 0 \\ N & 0 & 0 & \cdots & 0 & 0 \\ 0 & N & 0 & \cdots & 0 & 0 \\ \vdots & \vdots & \vdots & & \vdots & \vdots \\ 0 & 0 & 0 & \cdots & N & 0 \end{bmatrix},$$

thus $M^n = 0$. Hence if $N^s = 0$ and $n > s$,then $\sigma^*(I + tN) = 0$.

Proof.

$$\begin{aligned} \sigma^*(I + tN) &= [(I + M)(I - M + M^2 + \cdots + (-1)^{n-1}M^{n-1})((I + M) + A)] \\ &= [(I + M)(I + (I - M + M^2 + \cdots + (-1)^{n-1}M^{n-1})A)] \\ &= [I + M] + [I + (I - M + M^2 + \cdots + (-1)^{n-1}M^{n-1})A] \\ &= [I + (I - M + M^2 + \cdots + (-1)^{n-1}M^{n-1})A]. \end{aligned}$$

It's easy to see that $I + (I - M + M^2 + \cdots + (-1)^{n-1}M^{n-1})A$ is an upper triangular matrix with diagonal entries 1. Therefore $[I + (I - M + M^2 + \cdots + (-1)^{n-1}M^{n-1})A] = 0$ in $K_1(R[t^n])$. □

Step 4. Assume that $Nil(R)$ is finitely generated; i.e. $Nil(R) = C^m \oplus A$ with A finite. Let p be a prime number, $p > |A|$. Then $f\colon Nil(R) \to Nil(R)$ defined by $f(x) = px$ is clearly 1-1.

Step 5. It follows from Step 3 that if $Nil(R)$ is finitely generated, then there exists s_0 such that if $n \geq s_0$ then $\sigma^*(x) = 0$ for all $x \in Nil(R)$.

Step 6. Pick a prime number p with $p > \max\{s_0, |A|\}$. Let $n = p$ and $x \in Nil_n(R)$, then by Step 1, 2 and 5, $nx = \sigma^*(\sigma_*(x)) = 0$. Thus $x = 0$ by Step 4. This finishes the proof of Theorem 6.11.

Remark 6.12. The references for this section are [1] and [18].

6.2 Edwards-Cannon's example

In this section we give an example of a non-combinatorial (non-Piecewise Linear; written non PL) triangulation of a manifold.

Definition 6.13. *Let v be a vertex of a (finite) simplicial complex K. Its* star *$St(v)$ is the union of all the simplices of K having v as a vertex and its* link *$Lk(v)$ is the subcomplex of $St(v)$ consisting of all simplices which* don't *have v as a vertex.*

Example 6.14. Figure 6.1.

Note that $St(v)$ is the cone on $Lk(v)$, expressed $St(v) = cLk(v)$. Also notice that

$$\{St(v) - Lk(v) | v \quad \text{vertices of} \quad K\}$$

is an open cover of K; and hence if each $Lk(v)$ is homeomorphic to S^{n-1}, then K is a closed n-dimensional manifold.

This motivates the following

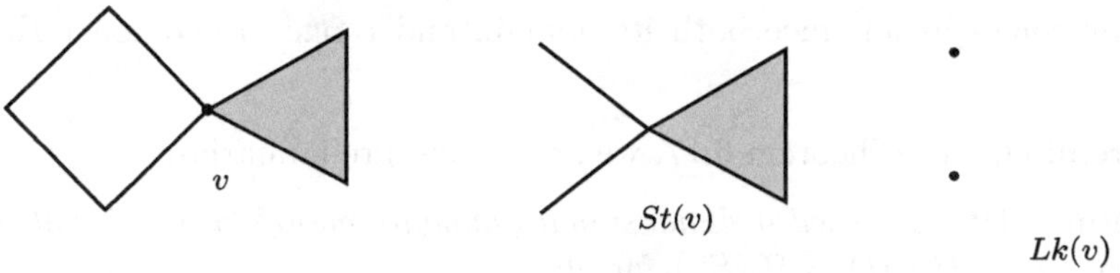

Figure 6.1

Definition 6.15. *A triangulation K of a closed n-dimensional manifold is (weakly)* combinatorial *if $Lk(v)$ is homeomorphic to S^{n-1} for each vertex v of K.*

Proposition 6.16. *Every smooth triangulation K of a closed smooth manifold M^n is weakly combinatorial.*

Theorem 6.17 (Edwards-Cannon). *There exists a triangulation of S^5 which is* not *weakly combinatorial.*

Addendum 6.18. *Every closed smooth manifold M^n $(n \geq 5)$ supports a triangulation which is* not *weakly combinatorial.*

Proof of Proposition 6.16. Let v be a vertex of K. Then $St(v)$ is a closed neighborhood of v in M^n. Let S_t^{n-1}, $t \in (0,1)$ be the family of concentric smooth codimension-1 spheres in M^n centered at v with respect to a smooth chart of M^n containing v. Choose ϵ small enough so that $S_\epsilon^{n-1} \subset St(v)$. Then each $x \in S_\epsilon^{n-1}$ lies on exactly one ray r in $St(v) = cLk(v)$ where $r = [v, y]$ is the line segment in $cLk(v)$ starting at v and ending at some point $y \in Lk(v)$. We define a continuous map $f\colon S_\epsilon^{n-1} \to Lk(v)$ by $f(x) = y$ (Figure 6.2).

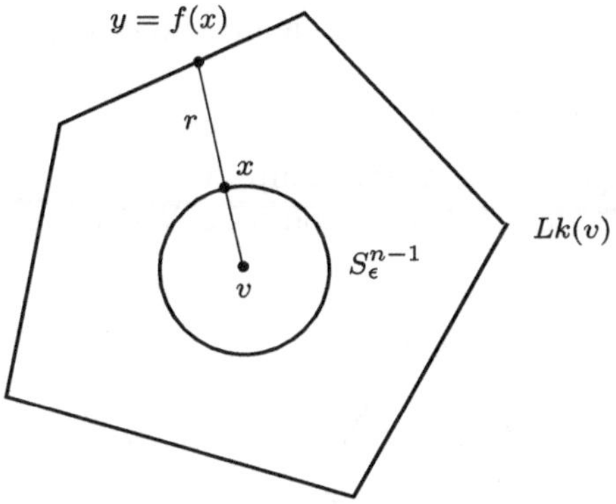

Figure 6.2

It is a straightforward consequence of the Jordan-Brouwer Separation Theorem that f is onto. And using the fact that K is a *smooth triangulation*, it is a differential topology exercise to show that f is 1-1 provided ϵ is sufficiently small. Hence

f is a homeomorphism since both its domain and range are compact Hausdorff spaces. □

Before discussing Theorem 6.17, we need some preliminaries.

Definition 6.19. *A closed n-dimensional (smooth) manifold M^n is called a* homology sphere *if $H_i(M) = H_i(S^n)$ for all i.*

Of course if $\pi_1(M^n) = 1$, then (as we have seen) M^n is homotopy equivalent to S^n, and hence homeomorphic to S^n by the solution to the Poincaré and Generalized Poincaré Conjectures. But in every dimension ≥ 3, there are non-simply connected homology spheres. The following 3-dimensional one $\mathscr{S}^3$ is of particular importance. It was constructed by Poincaré to give a counterexample to his original conjecture that "every homology 3-sphere is homeomorphic to S^3" and is called the *Poincaré homology sphere*.

To construct it consider the Platonic solid $\mathscr{D}$ — the dodecahedron. $\mathscr{D}$ has 12 faces and each face is a regular pentagon. The symmetry group of this solid is A_5 — the alternating group on 5 letters. It is a group of rotation of S^2; i.e. a subgroup of $SO(3) = \mathbb{R}\mathrm{P}^3$. The group of unit length quaternions S^3 is the universal covering group of $SO(3)$ and $p\colon S^3 \to SO(3)$ is 2-sheeted. Let $\overline{A}_5 = p^{-1}(A_5)$; it is a discrete *perfect* subgroup of S^3; i.e. $[\overline{A}_5, \overline{A}_5] = \overline{A}_5$ since A_5 is simple. The canonical quotient map

$$q\colon S^3 \to S^3/\overline{A}_5$$

is a 120-sheeted covering space, where $S^3/\overline{A}_5$ is the space of (left) cosets. The Poincaré homology sphere $\mathscr{S}^3$ is $S^3/\overline{A}_5$. It is clearly a closed 3-manifold and

$$H_1(\mathscr{S}^3) = \pi_1(\mathscr{S}^3)/[\pi_1(\mathscr{S}^3), \pi_1(\mathscr{S}^3)] = \overline{A}_5/[\overline{A}_5, \overline{A}_5] = 0.$$

And Poincaré duality gives that

$$H_2(\mathscr{S}^3) = H^1(\mathscr{S}^3) = \hom(H_1(\mathscr{S}^3), \mathbb{Z}) = 0.$$

Hence $\mathscr{S}^3$ is a homology 3-sphere.

Recall that the suspension ΣX of a compact space X is the union of two cones on X, c^+X and c^-X, along their common base X (Figure 6.3).

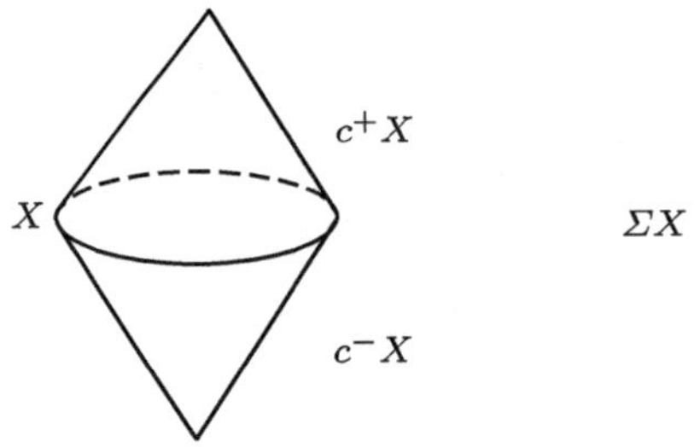

Figure 6.3

Van Kampen Theorem shows that $\pi_1(\Sigma X) = 1$ when X is arc connected. Hence $\pi_1(\Sigma M^n) = 1$ when M^n is a homology n-sphere and one easily sees, using the Mayer-Vietoris sequence, that $H_i(\Sigma M^n) = H_i(S^{n+1})$ for all i. But ΣM^n is *not* homeomorphic to S^{n+1} since ΣM^n is not a topological manifold. (The cone points do *not* have open neighborhoods homeomorphic to $\mathbb{R}^n$. This will be proven below.) On the other hand it was suspected in 1963 that its second suspension $\Sigma^2 M^n = \Sigma(\Sigma M^n)$ is a manifold and hence $\Sigma^2 M^n$ is homeomorphic to S^{n+2} by Smale's Generalized Poincaré Conjecture. In fact Milnor in 1963 (Annals of Math — Problems compiled by Lashof) listed this as one of the seven fundamental questions in manifold topology. It, along with the other six, have since been solved.

Theorem 6.20 (Edwards-Cannon Theorem). *The double suspension $\Sigma^2 M^n$ of any homology n-sphere M^n is homeomorphic to S^{n+2}.*

Remark 6.21. Edwards (1975) showed $\Sigma^3 M^n = S^{n+3}$ and Cannon (1977) extended this work showing $\Sigma^2 M^n = S^{n+2}$.

Now we return to the proof of Theorem 6.17. Let K be a smooth triangulation of the Poincaré 3-sphere $\mathscr{S}^3$. Then $T = \Sigma^2 K$ is a triangulation of $\Sigma^2 \mathscr{S}^3 = S^5$. We will show that T is *not* weakly combinatorial. Now

$$T = \Sigma \mathscr{K} = c^+ \mathscr{K} \cup_{\mathscr{K}} c^- \mathscr{K}$$

where $\mathscr{K} = \Sigma K$. Let c^+ be the cone point of $c^+\mathscr{K}$, then $Lk(c^+) = \mathscr{K}$. But $\mathscr{K}$ is *not* a sphere since it is homeomorphic to $\Sigma \mathscr{S}^3$ which is *not* a manifold, as was asserted above. So to complete our proof, it remains to verify this assertion.

Assertion 6.22. *Let M^n be an n-dimensional manifold ($n \geq 2$) such that $\pi_1(M) \neq 1$. Then $c \in cM^n$* does not *have an open neighborhood homeomorphic to $\mathbb{R}^{n+1}$.*

Proof. Our proof is by contradiction; i.e. we assume such a neighborhood U exists and show that this assumption leads to a contradiction. By our assumption we can identify a neighborhood V of c in cM^n with cS^n, both with the same cone point c. Recall that

$$cM^n = M^n \times [0,1]/\equiv$$

where $(x, 0) \equiv (y, 0)$ for all $x, y \in M^n$. Hence

$$M^n \times \epsilon \subset cS^n - c \subset cM^n - c \tag{6.4}$$

for some positive number ϵ which is sufficiently small. But the composite of the two inclusion maps in (6.4)

$$M^n \times \epsilon \subset cM^n - c$$

induces an isomorphism on π_1. However this composite factors through the simply-connected space

$$cS^n - c = S^n \times (0, 1].$$

This contradicts the fact that $\pi_1(M^n) \neq 1$. □

6.3 The Hauptvermutung

Hauptvermutung. *If a compact topological space can be triangulated, then triangulation is unique up to equivalence.*

Let B be a compact topological space and $\varphi\colon K \to B$, $\psi\colon L \to B$ be homeomorphisms where K and L are finite simplicial complexes.

Definition 6.23. *These triangulations are (rectilinearly)* equivalent *if they have isomorphic rectilinear subdivisions; i.e. there exists a finite complex P and homeomorphisms $\alpha\colon P \to K$, $\beta\colon P \to L$ such that for each simplex σ in P, there exists a simplex $\hat{\sigma}$, $\hat{\sigma}'$ of K, L respectively with $\alpha(\sigma) \subset \hat{\sigma}$, $\beta(\sigma) \subset \hat{\sigma}'$ and $\alpha|_\sigma$, $\beta|_\sigma$ are linear maps. (K,φ) and (L,ψ) are* strongly equivalent *if the rectilinear subdivisions $\alpha\colon P \to K$ and $\beta\colon P \to L$ can be chosen so that $\varphi \circ \alpha \simeq \psi \circ \beta$.*

Theorem 6.24 (Whitehead). *If M is a smooth manifold, then it has a smooth triangulation. And any two smooth triangulations are strongly equivalent.*

Milnor (1961) [16] gave the first counterexample to the conjecture (i.e. Hauptvermutung). In 1965, Stallings gave a very general class of counterexamples. In fact he showed the following

Theorem 6.25 (Stallings). *Let M^n be a closed smooth manifold such that $Wh(\pi_1 M) \neq 0$ and $n \geq 5$. Then ΣM^n has 2 triangulations which are* not *equivalent.*

Note that if ΣM is a manifold then M is simply-connected and $Wh(\pi_1 M) = 0$. So this left open the Hauptvermutung for manifolds. But this is really not the right question for manifolds since the Edwards-Cannon result (1976—1978) showed that every triangulable manifold of dimension≥ 5 has inequivalent triangulations. The more important question is whether a topological manifold M can have two smooth structures α, β where the smooth triangulations of α are *inequivalent* to those of β. This was solved in the *negative* by Siebenmann in 1969. We will discuss his solution to this problem. It can be developed into a whole theory called PL-structures on a manifold M^n. In particular if $H^3(M^n; \mathbb{Z}_2) = 0$ and $n \geq 5$, then M^n supports at most one equivalence class of smooth triangulations.

Definition 6.26. *A combinatorial structure on S^n is any triangulation equivalent to the boundary of an $(n+1)$-simplex. (Caveat: by Edwards-Cannon S^n has non-combinatorial triangulations when $n \geq 5$.) A triangulation $K \xrightarrow{\varphi} M^n$ is PL (piecewise linear) if for each vertex $v \in K$ the link of v, $Lk(v)$, is a combinatorial structure on S^{n-1}.*

Remark 6.27. By Edwards-Cannon (Theorem 6.17) S^n has non-combinatorial triangulation when $n \geq 5$.

To discuss Stallings' result, we need to return to talking about Whitehead torsion. In particular we will associate to any homotopy equivalence $f\colon K \to L$ between finite CW-complexes an element $\tau(f) \in Wh(\pi_1 L)$ called its torsion. If

$(W^{n+1}; M^n, N^n)$ is an h-cobordism, then $\tau(W)$ defined before in terms of a handlebody decomposition of W with only 2- and 3-handles turns out to be $\tau(\sigma)$ where $\sigma: M \to W$ is the inclusion and M, W are given any Whitehead triangulations; i.e. is independent of choices. This is proven in Milnor's long 1966 Bull. AMS article [15] and we will not go into much more detail other than making clear definitions. (Most of the details are due to Whitehead (1950).)

We will also discuss the fundamental question that $\tau(f)$, where $f: X \to Y$ is a homotopy equivalence between finitely triangulated spaces, is independent of the triangulations of X and Y. This was eventually proven to be true by Chapman (1974), completing earlier work of Kirby-Siebenmann. Perhaps the first result on this is due to Farrell-Hsiang (1967), giving the first example of an h-cobordism $(W; M, N)$ which is *not* homeomorphic to $M \times [0,1]$. We will give this example in §6.6.

6.4 Whitehead torsion

Let $K \overset{i}{\subset} L$ be a connected subcomplex of a connected finite simplicial complex L such that i is a homotopy equivalence and Γ denote $\pi_1(L)$, which we identify with the group of deck transformations of the universal cover of L, $\widetilde{L} \xrightarrow{p} L$. Note that $\widetilde{K} = p^{-1}(K) \to K$ is the universal cover of K. For each simplex σ of L fix a lift $\hat{\sigma}$ of it to $\widetilde{L}$. The simplicial chain group $C_n(\widetilde{L}, \widetilde{K})$ is a free $\mathbb{Z}\Gamma$-module with base $\{\hat{\sigma} \mid \sigma \text{ is an } n\text{-simplex in } L - K\}$ and $C(\widetilde{L}, \widetilde{K}) = \{C_n(\widetilde{L}, \widetilde{K})\}$ is an acyclic based chain complex over $\mathbb{Z}\Gamma$ and hence has a torsion $\tau(L, K) \in Wh(\Gamma)$. Note this element is independent of the choice of $\hat{\sigma}$ since $Wh(\Gamma) = K_1(\mathbb{Z}\Gamma)/\pm\Gamma$.

Remark 6.28. This definition is easily extended to handle the more general situation where K and L are finite CW-complexes. Also base points are unimportant since conjugation of an element of $GL_m(\mathbb{Z}\Gamma)$ by the diagonal matrix γI, $\gamma \in \Gamma$, induces the identity on $Wh(\Gamma) = K_1(\mathbb{Z}\Gamma)/\pm\Gamma$.

If $f: A \to B$ is a "cellular map" between finite CW-complexes A, B, which is a homotopy equivalence, we associate a torsion $\tau(f) \in Wh(\pi_1(B))$ as follows. The mapping cylinder $M_f = A \times [0,1] \cup_f B$, where $(a, 1) \in A \times [0,1]$ is identified to $f(a) \in B$, has a natural finite CW-structure: cells of $A = A \times 0$, cells of B and $\sigma \times (0,1)$ where σ is a cell of A.

Recall f is *cellular* means for each n-cell σ of A, $f(\sigma)$ is contained in the n-skeleton of B (i.e. the union of all i-cells of B, $i \leq n$). A is a subcomplex of M_f and the inclusion is a homotopy equivalence (if f is a homotopy equivalence). Therefore $\tau(M_f, A) \in Wh(\pi_1(M_f))$ is defined and $\pi_1(M_f)$ is naturally isomorphic to $\pi_1(B)$. Hence we define

$$\tau(f) = \tau(M_f, A) \in Wh(\pi_1(B)).$$

Remark 6.29. Also $B \subset M_f$ is a subcomplex which is canonically a deformation retract of M_f (just send $(x,t) \in A \times [0,1]$ to $f(x) \in B$). But $\tau(M_f, B) = 0$. It is a consequence of the following additivity formula and Addendum 6.30 below.

Fact 1. If $A \subset B \overset{j}{\subset} C$ are finite CW-complexes (and subcomplexes) each of them is a deformation retract of the next. Then

$$\tau(C, A) = \tau(C, B) + j_*\tau(B, A)$$

where $j_*: Wh(\pi_1(B)) \to Wh(\pi_1(C))$ is the isomorphism induced by the isomorphism $j_*: \pi_1(B) \to \pi_1(C)$.

Addendum 6.30. *Since M_f "collapses" to B, we can use Fact 1 to show, in a finite number of steps, that $\tau(M_f, B) = 0$. Namely, we can filter M_f by subcomplexes*

$$B = B_0 \subset \cdots \subset B_i \subset B_{i+1} \subset \cdots \subset B_n = M_f$$

such that $B_{i+1} = B_i \cup \sigma$ where σ is a closed cell and $\partial\sigma - B_i \cap \sigma$ is an (open) codimension-1 cell in σ. (This is the definition of "collapse".) (Or more generally, if $B_i \subset B_{i+1}$ is a homotopy equivalence for each i and $B_{i+1} = B_i \cup \sigma$, called a "generalized collapse".) It is easy to see that M_f has such a structure relative to B.

Now clearly $C_*(\widetilde{B}_{i+1}, \widetilde{B}_i) = C_*(B_{i+1}, B_i) \otimes_{\mathbb{Z}} \mathbb{Z}\pi_1(B_{i+1})$, hence

$$\tau(B_{i+1}, B_i) \in \operatorname{im}(Wh(1) \to Wh(\pi_1 B_{i+1}))$$

and hence is 0 since $Wh(1) = 0$.

If a homotopy equivalence $f: A \to B$ is not cellular, to define $\tau(f)$, first homotope f to a cellular map $\bar{f}$, and then define $\tau(f) = \tau(\bar{f})$. That $\tau(f)$ is well-defined is a consequence of the following

Fact 2. If homotopy equivalences f and $g: A \to B$ are homotopic cellular maps, then $\tau(f) = \tau(g)$.

This yields that, not only is $\tau(f)$ well-defined, but also

Corollary 6.31. *If $f, g: A \to B$ are homotopic homotopy equivalences, then $\tau(f) = \tau(g)$.*

And a consequence of Facts 1 and 2 is

Corollary 6.32. *If $f: A \to B$ and $g: B \to C$ are homotopy equivalences between finite CW-complexes, then $\tau(g \circ f) = \tau(g) + g_*\tau(f)$. Here $g_*: Wh(\pi_1 B) \to Wh(\pi_1 C)$.*

For a proof see [15, pp. 383–384] (Figure 6.4).
An immediate consequence of this corollary is

Addendum 6.33. *If $f: A \to B$ and $g: B \to A$ are homotopy inverses; i.e. $g \circ f \simeq \mathrm{id}_A$, $f \circ g \simeq \mathrm{id}_B$, where A and B are finite complexes, then "$\tau(f) = -\tau(g)$"after identifying $Wh(\pi_1 A)$ to $Wh(\pi_1 B)$ via the isomorphism f_*. More precisely, $\tau(f) = -f_*\tau(g)$.*

Proof. $\tau(f \circ g) = f_*\tau(g) + \tau(f)$ and $\tau(f \circ g) = \tau(\mathrm{id}_A) = 0$. This is because $M_{\mathrm{id}_A} = A \times [0,1]$ and clearly $(M_{\mathrm{id}_A}, A \times 1) = (M_{\mathrm{id}_A}, A \times 0)$. □

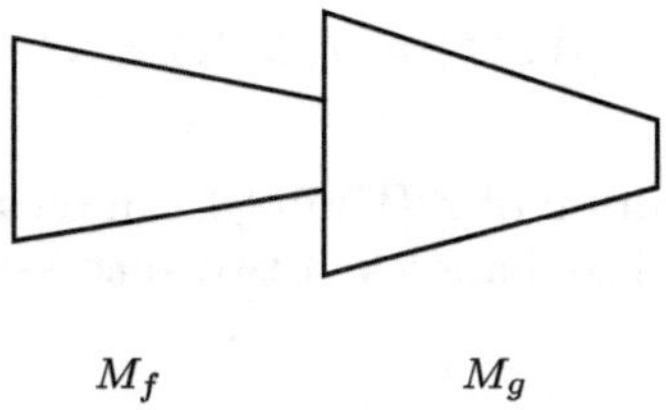

Figure 6.4

The following fundamental result is due to J. H. C. Whitehead. It's proof is a bit involved; but see [15, pp. 378–381] for an accessible proof.

Lemma 6.34. *If $f: A \to B$ is a "subdivision"; i.e. f is a homeomorphism and f^{-1} is cellular, then $\tau(f) = 0$.*

And it is easy to deduce the following (from Addendum 6.30 and Corollary 6.32)

Corollary 6.35. *Let $\sigma: K \subset L$ be a subcomplex of a finite CW-complex L and σ is a homotopy equivalence. Then $\tau(L, K) = \tau(\sigma)$.*

Note that Whitehead showed that any two smooth triangulations of a smooth manifold are *strongly equivalent.* Hence if $f: M \to N$ is a homotopy equivalence between compact smooth manifolds (possibly with boundaries), then $\tau(f) \in Wh(\pi_1 N)$ is well-defined. For this, we just use any smooth triangulations for M and N; i.e. consider

$$\begin{array}{ccc} M & \xrightarrow{f} & N \\ \varphi \uparrow & & \uparrow \psi \\ K & \xrightarrow{\psi^{-1} \circ f \circ \varphi} & L \end{array}$$

and let $\tau(f) = \psi_* \tau(\psi^{-1} \circ f \circ \varphi)$. This is a routine consequence of the definition of strong equivalence and the above corollaries.

Definition 6.36. A simple homotopy equivalence $f: K \to L$ *between finite CW-complexes is a homotopy equivalence such that $\tau(f) = 0$. If $f: M \to N$ is a homotopy equivalence where M, N are compact smooth manifolds, then f is called* simple *if for any (hence every) smooth triangulations K and L of M and N, respectively, $f: K \to L$ is simple.*

Since smooth triangulations are strongly equivalent, this concept is well-defined. Note that compositions of simple homotopy equivalences are again simply homotopy equivalence.

<u>h-cobordism torsion</u>

Let $(W; M, N)$ be a smooth h-cobordism, where $M = \partial^- W$. Then we define $\tau(W) = \tau(\sigma)$ where $\sigma: M \subset W$ is the inclusion. More precisely, let (L, K) be a smooth triangulation of (W, M).

Definition 6.37. $\tau(W) = \tau(W, M) = \tau(L, K) = \tau(\sigma)$, *where* $\sigma\colon K \subset L$ *is the inclusion.*

We gave a different definition of $\tau(W)$ in §3.1 in terms of a handlebody decomposition of W consisting of just handles of two successive indices s, $s+1$ (Figure 6.5).

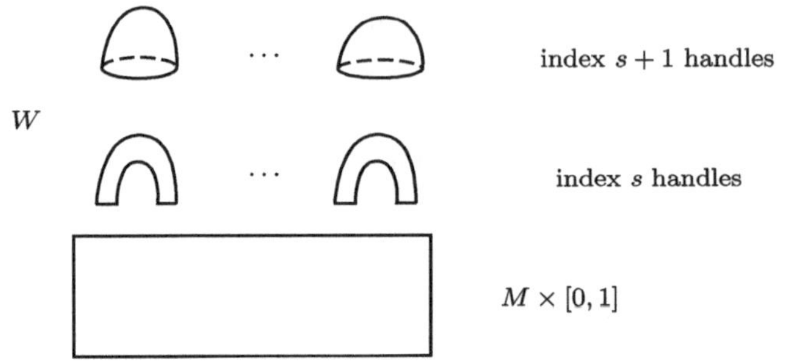

Figure 6.5

If A is the intersection matrix (with entries in $\mathbb{Z}\Gamma$, $\Gamma = \pi_1(W)$) giving how the index $s+1$ handles attach to index s handles, we have

Theorem 6.38. *The class of* A *in* $Wh(\Gamma)$ *equals* $(-1)^s\tau(W)$.

Remark 6.39. This is proven in [15, Theorem 9.3]. It depends on an algebraic result (algebraic subdivision Theorem 5.2, [15, p. 372]), and more basically on Whitehead smooth triangulation result.

Now we recall the

s-Cobordism Theorem 6.40 (Barden-Mazur-Stallings). *Let* M^m *be a closed smooth manifold* ($m \geq 5$) *and let* $\Gamma = \pi_1(M^m)$. *Then* $W \mapsto \tau(W) \in Wh(\Gamma)$ *gives a bijective correspondence between diffeomorphism classes (relative to* M*) of smooth h-cobordisms* W *with* $\partial^- W = M$ *and elements in* $Wh(\Gamma)$, *with* $W = M \times [0,1]$ *corresponding to* 0.

Duality Theorem

Let $\omega\colon \Gamma \to C_2$ be a homomorphism. Using ω we define an involution $-\colon Wh(\Gamma) \to Wh(\Gamma)$ as follows. First define an anti-involution $a \mapsto \bar{a}$ of the group ring $\mathbb{Z}\Gamma$ by

$$a = \sum_{g\in\Gamma} n_g g \mapsto \bar{a} = \sum_{g\in\Gamma} \omega(g) n_g g^{-1}.$$

(The most important case is where $\omega \equiv 1$, then $\sum_{g\in\Gamma} n_g g \mapsto \sum_{g\in\Gamma} n_g g^{-1}$.) This induces an anti-involution of $GL(\mathbb{Z}\Gamma)$ given by $A = (a_{ij}) \mapsto A^* = (\bar{a}_{ji})$. This induces the involution $x \mapsto \bar{x}$ on the abelianization $K_1(\mathbb{Z}\Gamma)$ and hence on $Wh(\Gamma) = K_1(\mathbb{Z}\Gamma)/\pm\Gamma$.

Duality Theorem 6.41. *$\tau(-W^n) = (-1)^{n+1}\overline{\tau(W^n)}$, relative to $\omega\colon \pi_1(W) \to H_1(W) \to \mathbb{Z}_2$, which is the 1st Stiefel-Whitney class of W; i.e. for each loop $g\colon S^1 \to W$, $\omega(g) = 1$ or -1 depending on whether the pullback bundle g^*TW is trivial or not.*

Remark 6.42. Here $-W$ denotes "W turned upside down"; i.e. $\partial^-(-W) = \partial^+ W$.

Proof. Consider the codimension-1 submanifold of W^n, M_s^{n-1}, which separates the s- and $(s+1)$- handles. Let $\mathscr{S}_1, \cdots, \mathscr{S}_m$ be the attaching spheres for the $(s+1)$-handles (each $\mathscr{S}_i$ is diffeomorphic to S^s) and $S_1, \cdots, S_m$ be the complementary spheres for the s-handles (each S_i is diffeomorphic to S^{n-1-s}). For the dual handle decomposition of $-W$ which consists of $(n-s)$ and $(n-(s+1))$-handles, M_s is still the separating membrane and $S_1, \cdots, S_m$ are now the attaching spheres; while $\mathscr{S}_1, \cdots, \mathscr{S}_m$ are the complementary spheres. Let $*$ be the base point in M_s, a_i, b_j be base points in $\mathscr{S}_i$, S_j respectively and α_i, β_j be curves in M_s connecting $*$ to a_i, b_j respectively. And consider the following picture, in which x is a point in the transversal intersection $\mathscr{S}_i \cap S_j$ (Figure 6.6).

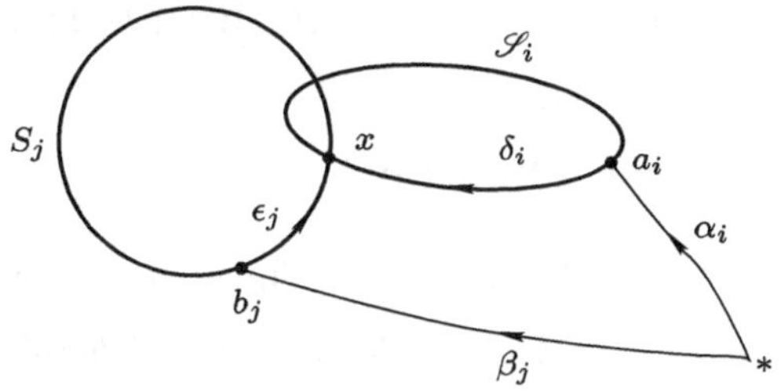

Figure 6.6

We also fix orientations for $\mathscr{S}_i$, S_j and T_*M_s. Recall the attaching $m \times m$ matrix A for W is defined by

$$A_{ij} = \sum_{x \in \mathscr{S}_i \cap S_j} a_x \gamma_x$$

where $a_x \in \mathbb{Z}$, $\gamma_x = \alpha_i \delta_i \varepsilon_j^{-1} \beta_j^{-1} \in \pi_1(M_s) = \pi_1(W) = \Gamma$ and the intersection $\mathscr{S}_i \cap S_j$ (in that order) gives an orientation to $T_x M_s$ which we compare to the one at $T_* M_s$ by translating along the arc $\alpha_i \delta_i$. If they agree $a_x = 1$; if not $a_x = -1$.

Now the attaching $m \times m$ matrix B for $-W$ can be "read off" from the same date; i.e.

$$B_{ji} = \sum_{x \in S_j \cap \mathscr{S}_i} b_x \gamma'_x.$$

As a set $\gamma'_x = \gamma_x$; but it is traveled in the opposite direction; hence $\gamma'_x = \gamma_x^{-1}$. Also the orientation at $T_x M_s$ for $-W$ is $(-1)^{s((n-1)-s)}$ times that of $T_x M_s$ for W since our order now is $S_j \cap \mathscr{S}_i$. Also to determine b_x we translate it back to $T_* M_s$ along $\varepsilon_j^{-1} \beta_j^{-1}$ instead of $\delta_i^{-1} \alpha_i^{-1}$. Hence $b_x = \omega(\gamma_x)(-1)^{s((n-1)-s)} a_x$. Therefore

$$\begin{aligned} B_{ji} &= (-1)^{s((n-1)-s)} \sum_{x \in S_j \cap \mathscr{S}_i} \omega(\gamma_x) a_x \gamma_x^{-1} \\ &= (-1)^{s(n-1)-s)} \overline{A}_{ij} \end{aligned}$$

Hence $B = (-1)^{s((n-1)-s)} A^*$, and $[B] = [A^*]$ in $Wh(\Gamma)$ since $Wh(\Gamma) = K_1(\mathbb{Z}\Gamma)/\pm\Gamma$. But $\tau(W) = (-1)^s[A]$ while $\tau(-W) = (-1)^{n-(s+1)}[B]$, therefore

$$\tau(-W) = (-1)^{n-(s+1)}[A^*] = (-1)^{n-(s+1)}(-1)^s\overline{\tau(W)} = (-1)^{n+1}\overline{\tau(W)}. \qquad \square$$

Corollary 6.43. *Let $(W^n; M, N)$ be a smooth h-cobordism ($n \geq 6$). And let $f\colon M \to N$ denote the homotopy equivalence which is the composite of the inclusion $\sigma\colon M \subset W$ and the deformation retraction $r\colon W \to N$. Then $\tau(f) = r_*(\tau(W) + (-1)^n\overline{\tau(W)})$.*

Proof. $\tau(f) = \tau(r \circ \sigma) = r_*\tau(\sigma) + \tau(r) = r_*\tau(W) + \tau(r)$. Let $i\colon N \subset W$ denote the inclusion. Then $\tau(i) = \tau(-W)$. But $r_*(\tau(i)) = -\tau(r)$ by Addendum 6.33. Therefore

$$\tau(f) = r_*(\tau(W) - \tau(-W)) = r_*(\tau(W) + (-1)^n\overline{\tau(W)}). \qquad \square$$

6.5 Proof of Stallings' Theorem

Now we are able to give a proof of Stallings' Theorem (Theorem 6.25).

Lemma 6.44 (Stallings). *Let $(W^n; M, N)$ be a smooth h-cobordism. Then $W - N$ is diffeomorphic to $M \times [0, 1)$.*

Proof. Let $x = \tau(W) \in Wh(\Gamma)$ where $\Gamma = \pi_1(W)$. By the s-Cobordism Theorem, there exists another h-cobordism $\mathscr{W}$ with $\partial^-\mathscr{W} = N$ and $\tau(\mathscr{W}) = -x$. Now glue W to $\mathscr{W}$ along N, the result $W \cup_N \mathscr{W}$ is an h-cobordism (Figure 6.7).

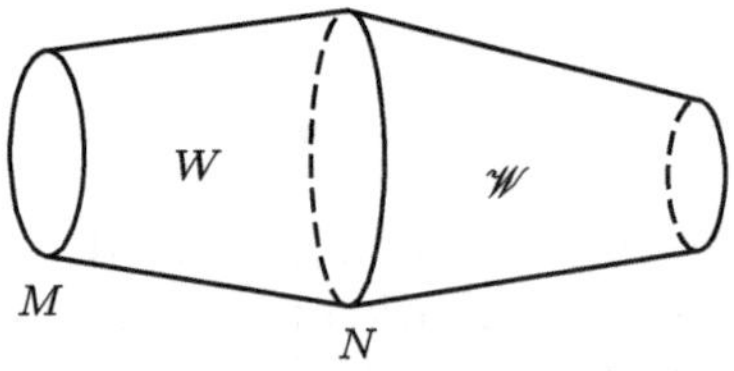

Figure 6.7

But by Fact 1,

$$\tau(W \cup_N \mathscr{W}) = \tau(W, M) + \tau(\mathscr{W}, N) = x - x = 0.$$

Therefore $W \cup_N \mathscr{W} = M \times [0, 1]$. Note this implies $\partial^+\mathscr{W} = M$. Thus we may glue $\mathscr{W}$ and W along M. The torsion of $\mathscr{W} \cup_M W$ is

$$\tau(\mathscr{W} \cup_M W) = \tau(\mathscr{W}, N) + \tau(W, M) = -x + x = 0.$$

Hence $\mathscr{W} \cup_M W = N \times [0,1]$. Consider Figure 6.8.

Thus we have $W \cup_N (N \times [0,\infty)) \cong M \times [0,\infty)$. But $W \cup_N (N \times [0,\infty)) \cong W - N$. This proves the lemma. □

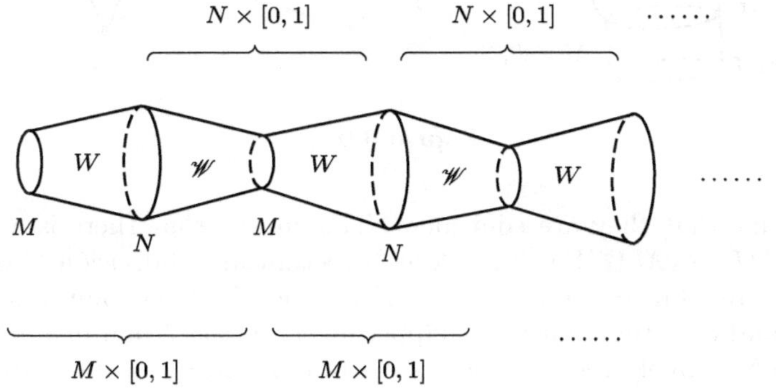

Figure 6.8

Corollary 6.45. $W - \partial W \cong M \times \mathbb{R} \cong N \times \mathbb{R}$.

Proof. $W - \partial W = M \times [0,\infty) - M \times \{0\} \cong M \times \mathbb{R}$ by the lemma. And replacing W by $-W$ in the above, we get $W - \partial W = -W - \partial(-W) = N \times [0,\infty) - N \times \{0\} \cong N \times \mathbb{R}$. □

Corollary 6.46. *If $(W; M, N)$ is an h-cobordism, then ΣM is homeomorphic to ΣN.*

Proof. $M \times \mathbb{R}$ is diffeomorphic to $N \times \mathbb{R}$ by the above corollary. But the 2 points compactification of $M \times \mathbb{R}$ is $M \times \mathbb{R} \cup \{\pm\infty\} = \Sigma M$. Likewise the 2 points compactification of $N \times \mathbb{R}$ is ΣN. □

Proof of Theorem 6.25. Let $x \in Wh(\pi_1 M)$ be a non-zero element, and $(W^{n+1}; M, N)$ be an h-cobordism with $\tau(W) = x$. Put two triangulations θ, ω on ΣM as follows:

Let K be a smooth triangulation of M, then $\theta = \Sigma K$. Let L be a smooth triangulation of W, inducing smooth triangulations R^- and R^+ of M and N respectively. Then ω is $cR^- \cup L \cup cR^+$, where $cM \cup W \cup cN$ is identified with ΣM as follows (Figure 6.9):

Write $\Sigma M = c^- M \cup c^+ M$. Identifying cM with $c^- M$. Then identifying $W \cup cN$ with W/N, which is the one point compactification of $W - N$, which is also the one point compactification of $M \times [0, +\infty)$, which is $c^+ M$.

The proof of this theorem is completed by the following assertion. □

Assertion 6.47. *θ and ω are inequivalent triangulations of ΣM.*

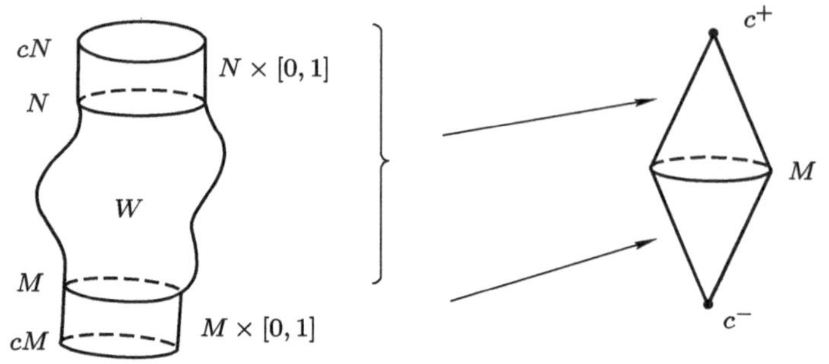

Figure 6.9

Proof. Assume that they are equivalent. This means that there is a homeomorphism $F\colon \Sigma M \to cM \cup W \cup cN$ taking a rectilinear subdivision θ' of θ isomorphically to a rectilinear subdivision ω' of ω. Since ΣM has only 2 non-manifold points (c^+ and c^-), they must be mapped by F to the 2 non-manifold points of $cM \cup W \cup cN$, namely the cone points c_M of cM and c_N of cN. After renaming things, we may assume that $F(c^+) = c_N$ and $F(c^-) = c_M$.

Now delete the open stars of c^-, c^+ and c_N, c_M (with respect to θ' and ω') from ΣM and $cM \cup W \cup cN$, respectively, and denote the resulting subcomplexes by A and B (Figure 6.10).

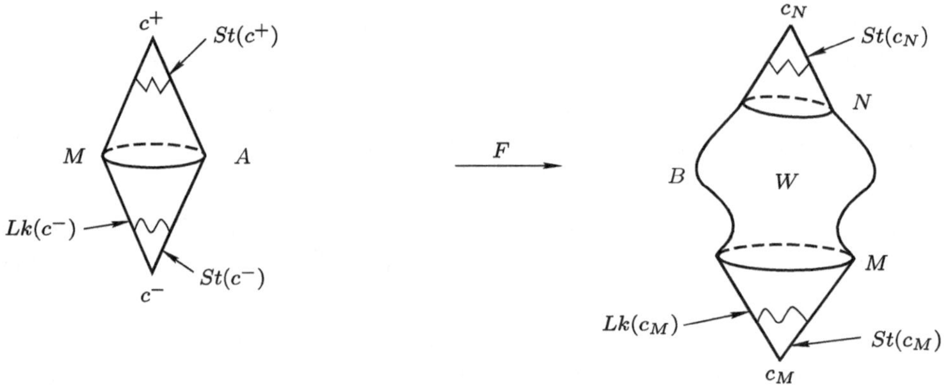

Figure 6.10

Then $F\colon A \to B$ is a simplicial isomorphism and hence $\tau(F) = 0$. Let σ_A and σ_B denote the inclusions of $Lk(c^-)$ and $Lk(c_M)$ into A and B, respectively. And let $f = F|_{Lk(c^-)}$. Since $F \circ \sigma_A = \sigma_B \circ f$, we have

$$\tau(F) + F_*(\tau(\sigma_A)) = \tau(\sigma_B) + (\sigma_B)_*(\tau(f)).$$

Since f is also a simplicial isomorphism, $\tau(f) = 0$, and hence

$$F_*(\tau(\sigma_A)) = \tau(\sigma_B). \tag{6.5}$$

The theorem is established after we have shown the following 2 facts:

1. $\tau(\sigma_A) = 0$;
2. $\tau(\sigma_B) = \tau(W) \neq 0$.

Since these obviously contradict Eq. (6.5).

Now we proceed to prove the above 2 facts. First, by applying Fact 1 and Corollary 6.35, we get that

$$\tau(\sigma_A) = \tau(A^-, Lk(c^-)) + \tau(A^+, M),$$

where $A^+ = c^+M$ – open star(c^+), $A^- = c^-M$ – open star(c^-) and the equation is understood in $Wh(\pi_1(A))$ by using the obvious group isomorphisms. So it suffices to show that both $\tau(A^-, Lk(c^-))$ and $\tau(A^+, M)$ are 0. Consider the pair $(A^-, Lk(c^-))$ first. Put a new *regular cell* structure θ'' on A^- by declaring cells of $M \subset A^-$ and $Lk(c^-) \subset A^-$ to be subcomplexes with an additional $(i+1)$-dimensional cell $\hat{\sigma}$ for each i-cell σ of M; namely $\hat{\sigma} = c^-\sigma$ – open star(c^-), where the star is taken in θ' (Figure 6.11).

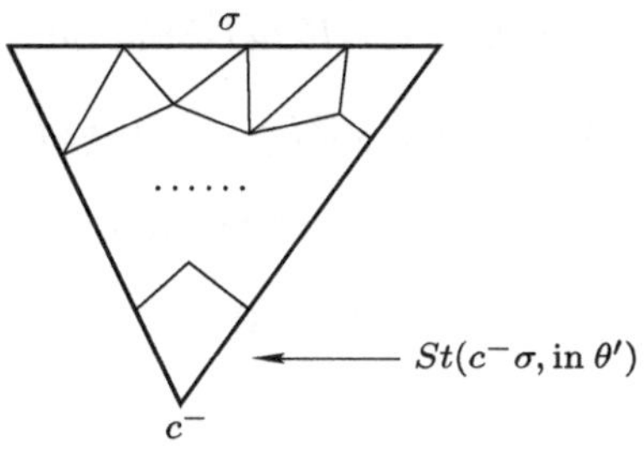

Figure 6.11

And note that θ' subdivides θ''. Hence by Lemma 6.34,

$$\tau_{\theta'}(A^-, Lk(c^-)) = \tau_{\theta''}(A^-, Lk(c^-)).$$

But the latter is zero since A^- collapses to $Lk(c^-)$ in θ''. A similar argument shows $\tau(A^+, M) = 0$; the argument corresponds to showing that $\tau_{\theta''}(A^-, M) = 0$. But this is true since A^- "generalized collapses" to M (see Addendum 6.30).

The proof of the second statement proceeds similarly. Again by Fact 1 and Corollary 6.35,

$$\tau(\sigma_B) = \tau(B^-, Lk(c_M)) + \tau(W, M) + \tau(B^+, N)$$

where $B^+ = cN$ – open star(c_N) and $B^- = cM$ – open star(c_M). Then by the same argument as above, one shows that $\tau(B^-, Lk(c_M)) = 0 = \tau(B^+, N)$. This completes the proof of the assertion. □

6.6 Farrell-Hsiang's example

Our next project is to prove the following

Theorem 6.48 (Farrell-Hsiang, Bull. AMS 1967). *There exists a smooth h-cobordism W^{n+2} for every $n \geq 5$, such that W is* not *topologically trivial; i.e. W is* not homeomorphic *to $\partial^- W \times [0,1]$.*

Remark 6.49. This result is now obsolete due to the later fundamental work of Kirby-Siebenmann (1969) where they proved that a smooth h-cobordism $(W^{m+1}; M^m, N^m)$, $m \geq 5$, is topologically a product (i.e. homeomorphic to $M^m \times [0,1]$) if and only if it is smoothly a product. And by the s-Cobordism Theorem, that implies $\tau(W) = 0$. Hence if $\tau(W) \neq 0$, it cannot be a topological product.

In this example, $\partial^- W$ is any manifold of the form $M^n \times S^1$ where M^n is a closed smooth manifold such that $Nil(\mathbb{Z}[\pi_1(M)]) \neq 0$; e.g. any manifold M^n such that $\pi_1(M) = \Gamma = C \times C_{p^2}$ (p a prime).

The following algebraic lemma will be needed in the proof of the above theorem.

Algebraic Lemma 6.50. *In the Bass-Heller-Swan formula*

$$Wh(\Gamma \times C) = Wh(\Gamma) \oplus \widetilde{K}_0(\mathbb{Z}\Gamma) \oplus Nil^+(\mathbb{Z}\Gamma) \oplus Nil^-(\mathbb{Z}\Gamma)$$

the two copies of $Nil(\mathbb{Z}\Gamma)$ are interchanged by the involution

$$-\colon Wh(\Gamma \times C) \to Wh(\Gamma \times C);$$

i.e. $\overline{Nil^+(\mathbb{Z}\Gamma)} = Nil^-(\mathbb{Z}\Gamma)$.

Proof. Let the invertible matrix $I + tN$ represent a typical element $x \in Nil^+(\mathbb{Z}\Gamma)$ where N is an $m \times m$ nilpotent matrix over $\mathbb{Z}\Gamma$ and t is a fixed generator for the infinite cyclic group C. Then $\overline{x}$ in $Wh(\Gamma \times C)$ is represented by

$$(I + tN)^* = I^* + \omega(t)t^{-1}N^* = I + t^{-1}M$$

where $M = \omega(t)N^*$. Since M is clearly nilpotent, $\overline{x} \in Nil^-(\mathbb{Z}\Gamma)$. □

Proof of Theorem 6.48. Let M^n be a closed smooth manifold with $\pi_1(M) = \Gamma$ such that $Nil(\mathbb{Z}\Gamma) \neq 0$. Then pick a non-zero element $x \in Nil(\mathbb{Z}\Gamma)$ and by the Bass-Heller-Swan formula, identify it to an element in $Wh(\Gamma \times C)$ through identifying $Nil(\mathbb{Z}\Gamma)$ with the summand $Nil^+(\mathbb{Z}\Gamma)$ in $Wh(\Gamma \times C)$. Then use the s-Cobordism Theorem to get the h-cobordism W^{n+2} with $\partial^- W = M \times S^1$, such that $\tau(W) = x$. We claim that with this choice W is *not homeomorphic* to $\partial^- W \times [0,1]$; i.e. W is not homeomorphic to $M \times S^1 \times [0,1]$.

Assume that W is homeomorphic to $M \times S^1 \times [0,1]$, we will produce a contradiction. Let $F\colon W \to M \times S^1 \times [0,1]$ be a homeomorphism such that $F(\partial^- W) = M \times S^1$. Let $f\colon \partial^- W \to M \times S^1$ be $F|_{\partial^- W}$ and $g\colon \partial^+ W \to M \times S^1$ be $F|_{\partial^+ W}$. Then

$$\varphi = g^{-1} \circ f\colon \partial^- W \to \partial^+ W$$

is a homeomorphism. Then an elementary argument shows that φ is homotopy equivalent to $r \circ \sigma$, where $\sigma\colon \partial^- W \to W$ is the inclusion and $r\colon W \to \partial^+ W$ is the deformation retraction. Hence $\tau(\varphi) = \tau(r \circ \sigma) = r_*(x + (-1)^n \overline{x})$ by Corollary 6.43. And by the Algebraic Lemma 6.50, $\overline{x} \in Nil^-(\mathbb{Z}\Gamma)$; i.e.

$$\tau(\varphi^{-1}) = -x + (-1)^{n+1}\overline{x} \in Nil^+(\mathbb{Z}\Gamma) \oplus Nil^-(\mathbb{Z}\Gamma) \subset Wh(\Gamma \times C),$$

and the $Wh(\Gamma)$ and $\widetilde{K}_0(\mathbb{Z}\Gamma)$ components of $\tau(\varphi^{-1})$ are both 0.

In the Ph. D. thesis of Farrell (Indiana J., 1971) it is shown that given a homotopy equivalence $\psi\colon N^{n+1} \to M^n \times S^1$ $(n \geq 5)$, where N and M are closed smooth manifolds, ψ can be "split" along $M \times 1$ if and only if $\tau(\psi) \in Wh(\Gamma) \subset Wh(\Gamma \times C)$. Here "split" means there exists a closed smooth submanifold $\mathscr{M}^n \subset N^{n+1}$ homotopy equivalent to M, with trivial normal bundle (i.e. $\nu_{\mathscr{M}} = \mathscr{M} \times [-1,1] \subset N^{n+1}$) so that the composition

$$N^{n+1} \to \nu_{\mathscr{M}}/\partial = (\mathscr{M} \times [-1,1])/\partial \to [-1,1]/\partial = S^1$$

is homotopic to $N \xrightarrow{\psi} M \times S^1 \to S^1$ (Figure 6.12).

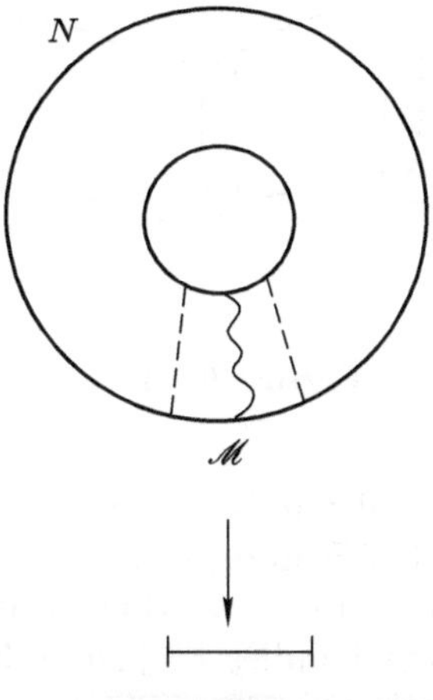

Figure 6.12

Hence φ^{-1} *cannot* be split along $M \times 1$ since $\tau(\varphi^{-1}) \notin Wh(\Gamma)$.

Moreover Farrell also showed that "some finite sheeted cover of ψ" splits if and only if the $\widetilde{K}_0(\mathbb{Z}\Gamma)$ component of $\tau(\psi)$ is 0.

The covers here being referred to are those corresponding to $\Gamma \times sC \subset \Gamma \times C = \pi_1(M \times S^1)$, i.e.

$$\begin{array}{ccc} N_s & \xrightarrow{\psi_s} & M \times S^1 \\ \downarrow & & \downarrow \scriptstyle{\mathrm{id} \times z^s} \\ N & \xrightarrow{\psi} & M \times S^1 \end{array}$$

Remark 6.51. The second fact is an easy consequence of the Novikov-Siebenmann Splitting Theorem and the geometric interpretation of the Bass-Heller-Swan Theorem given by Hsiang-Farrell (Bull. AMS 1967).

Let $N = \partial^+ W$ and applying this to $\psi = \varphi^{-1}$, we see that $(\varphi^{-1})_s$ splits for some $s > 1$. And consequently $\hat{\varphi}^{-1}\colon \hat{N} \to M \times \mathbb{R}$ "splits", where

$$\begin{array}{ccc} \hat{N} & \xrightarrow{\hat{\varphi}^{-1}} & M \times \mathbb{R} \\ \downarrow & & \downarrow{\scriptstyle \mathrm{id} \times \exp} \\ N & \xrightarrow{\varphi^{-1}} & M \times S^1 \end{array}$$

is a lift of φ^{-1} to the infinite cyclic cover $\hat{N}$ of N, which corresponds to $\Gamma \subset \Gamma \times C$.

"Splits" means that $\hat{N}$ contains a codimension-1 closed smooth submanifold $\mathscr{M}$ homotopy equivalent to M, whose complement $N - \mathscr{M}$ has exactly 2 components $\hat{N}_{-\infty}$ and $\hat{N}_{+\infty}$ each of which is unbounded; i.e. $\hat{N}_{-\infty} \cup \mathscr{M}$ and $\hat{N}_{+\infty} \cup \mathscr{M}$ are both not compact (Figure 6.13).

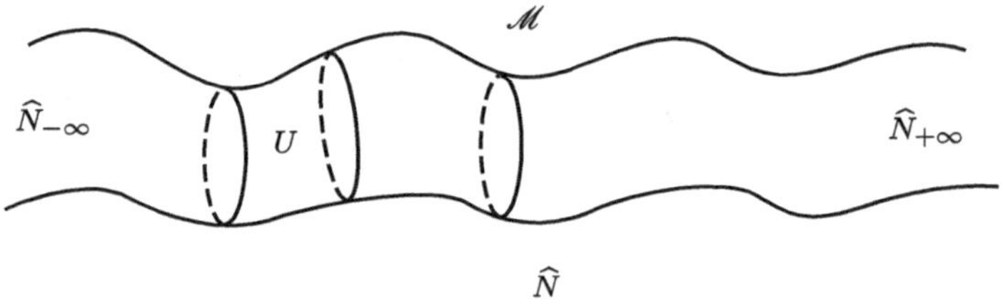

Figure 6.13

Let ε be a small positive real number, then $U = \hat{\varphi}(M \times (-\varepsilon, \varepsilon))$ is an open subset of $\hat{N}$ which is mapped diffeomorphically by $\hat{N} \to N$ to an open subset V of N. Now by an "engulfing" argument, due to Stallings, (using a preliminary handle decomposition involving handles of just indices 2 and 3, resulting from the Novikov-Siebenmann splitting argument,) one can find a self diffeomorphism $f\colon \hat{N} \to \hat{N}$ isotopic to $\mathrm{id}_{\hat{N}}$ such that $f(\mathscr{M}) \subset U$.

But then the image of $f(\mathscr{M})$ in N under $\hat{N} \to N$ splits $\varphi^{-1}\colon N \to M \times S^1$ (Figure 6.14).

This contradiction proves Theorem 6.48. □

Remark 6.52. The "engulfing" is done as follows. Using some preliminary construction found in the proof of the Novikov-Siebenmann Splitting Theorem, one can find a compact smooth codimension-0 cobordism W inside of $\hat{N}$ such that $\partial^+ W = M$ and $\partial^- W$ lies "weakly to the left of U" and so that $(W, \partial^- W)$ has a handlebody decomposition consisting of m handles of indices 2 and 3. "Weakly to the left of U" means

$$\partial^- W \subset \hat{\varphi}(M \times (-\infty, \varepsilon)).$$

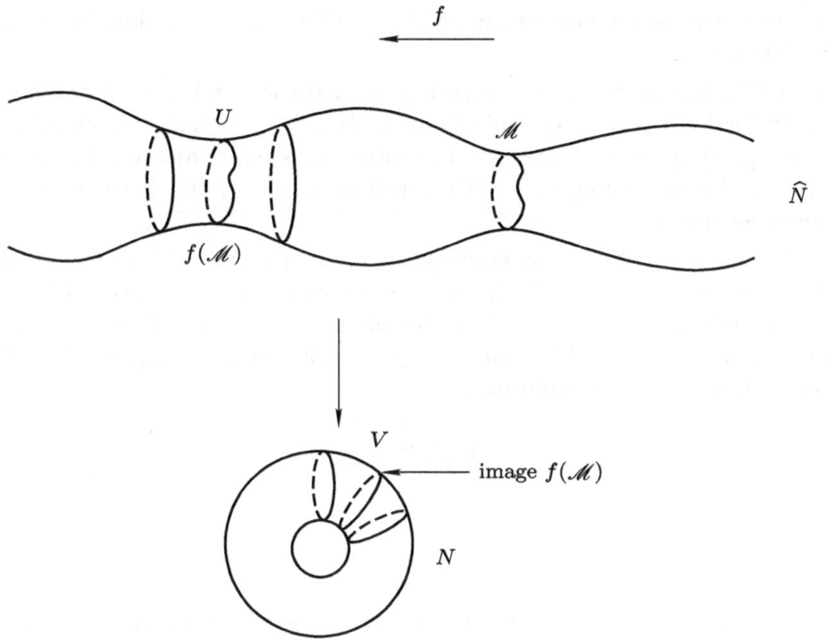

Figure 6.14

Now using the Whitney Embedding Theorem, we can isotope the cores of these handles (one at a time) (weakly) to the left of U. Then using the isotopy extension theorem, we can, in fact, isotope these handles ambiently into $\hat{\varphi}(M \times (-\infty, \varepsilon))$ in m steps. In particular $\mathscr{M}$ has been moved weakly to the left of U. Now repeat the same argument to the weak right of U, but move only within $\hat{\varphi}(M \times (-\infty, \varepsilon))$. The result of these two movements is the posited self-diffeomorphism f.

Note that if we had already started with $\mathscr{M}$ to the left of U (which could easily be arranged) then only the second of the two movements is needed to engulf $\mathscr{M}$ by U.

6.7 The structure set

Our final project is to give Siebenmann's counterexample to the manifold Hauptvermutung and more generally give some indication of the Kirby-Siebenmann smoothing theory. These results depend on calculating the number of closed smooth manifolds homotopy equivalent to the n-torus $T^n = S^1 \times \cdots \times S^1$ (n-factors).

For this we first make some definitions; i.e. we define the homotopy-smoothing structure set $\mathscr{S}(N^n)$, where N^n is a closed smooth manifold. It will be convenient to make the following assumption concerning $\Gamma = \pi_1(N^n)$; namely that it is "K-flat"; i.e. $Wh(\Gamma \times C^m) = 0$ for all $m \geq 0$, where $C^m = C \oplus \cdots \oplus C$ (m-summands) is the free abelian group of rank m. By the Bass-Heller-Swan formula free abelian groups are K-flat and hence we can let $N^n = T^n$ in our discussion; in fact, this is the crucial case.

Remark 6.53. Lowell Jones and Farrell proved (in the Asian J. of Math (1998)) that every torsion-free discrete subgroup of $GL_n(\mathbb{R})$ is K-flat; this includes many of the groups of geometric interest. The basic unsolved conjecture in this area is "Γ is K-flat if it is torsion-free." "Torsion-free" means that 1 is the only element of finite order in Γ.

An element in $\mathscr{S}(N^n)$ is an equivalence class of pairs (M^n, f) where M^n is a closed smooth manifold and $f\colon M \to N$ is a homotopy equivalence. We say that (M_1, f_1) is equivalent to (M_2, f_2) (denoted as $(M_1, f_1) \equiv (M_2, f_2)$) if there exists a diffeomorphism $F\colon M_1 \to M_2$ such that $f_2 \circ F$ is homotopic to f_1; i.e. the following diagram commutes up to homotopy

$$\begin{array}{ccc} M_1 & \xrightarrow{f_1} & N \\ \downarrow{\scriptstyle F} & \nearrow{\scriptstyle f_2} & \\ M_2 & & \end{array}$$

Then $\mathscr{S}(N^n)$ is a "pointed set"; i.e. it has a natural base point, namely the class of (N^n, id). It is important to generalize this definition to include the case where N^n is a compact smooth manifold with possibly non-empty boundary ∂N. Then in the pair (M^n, f) we require M compact and that $f|_{\partial M}\colon \partial M \to \partial N$ is a diffeomorphism. Furthermore for $(M_1, f_1) \equiv (M_2, f_2)$ we require that $f_2 \circ F|_{\partial M_1} = f_1|_{\partial M_1}$ and the homotopy $f_2 \circ F \simeq f_1$ is constant on ∂M_1. The following result is implicit in (Burghelea, Topology, 1977).

Theorem 6.54. *For $n \geq 5$, assume that $\pi_1(N)$ is K-flat, then*

$$0 \to \mathscr{S}(N^n \times I) \xrightarrow{\beta} \mathscr{S}(N^n \times S^1) \xrightarrow{\alpha} \mathscr{S}(N^n) \to 0$$

is a split short exact sequence of pointed sets, with a splitting $\gamma\colon \mathscr{S}(N^n) \to \mathscr{S}(N^n \times S^1)$.

Remark 6.55. $\mathscr{S}(N \times I)$ has a group structure gotten by concatenation; see Figure 6.15. And the group structure is abelian if $N = K \times I$. So if $N = M^{n-2} \times I^2$, then all three sets above are abelian groups and

Addendum 6.56. $\mathscr{S}(N \times S^1) = \mathscr{S}(N \times I) \times \mathscr{S}(N)$ *as groups, since β, α and γ are group homomorphisms.*

Proof of the Theorem. We do the case where N is closed. Let $f\colon M \to N \times S^1$ represent $x \in \mathscr{S}(N \times S^1)$. Then fibering Theorem 4.29 applies to give $N_1 \subset M$, where $\hat{f}\colon M \to N \times S^1$ is homotopic to f such that $\hat{f}|_{N_1}\colon N_1 \to N$ is a homotopy equivalence (Figure 6.16).

Then $\alpha(x)$ is taken to be the equivalence class of $(N_1, \hat{f}|_{N_1})$. Another application of the fibering theorem shows that α is well-defined (i.e. independent of the choices of (M, f) and $\hat{f}$).

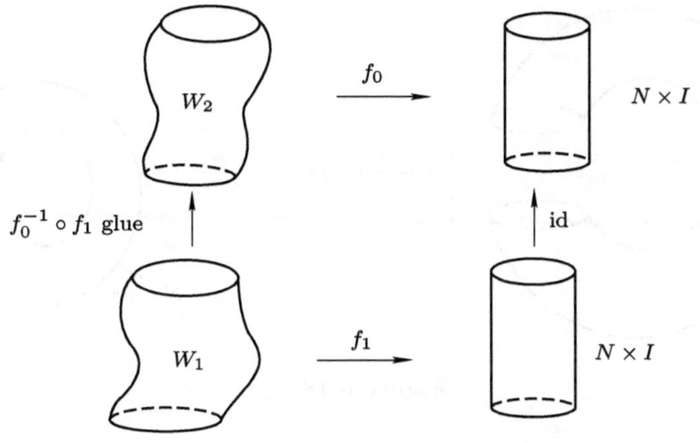

Figure 6.15

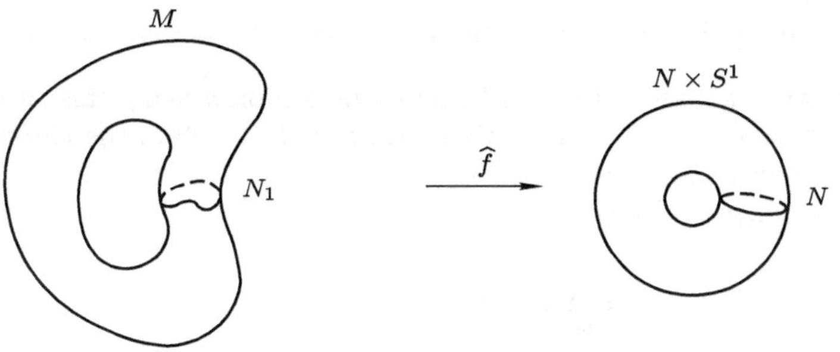

Figure 6.16

The splitting γ is given as follows: let (M, φ) represent an element $y \in \mathscr{S}(N)$, then $(M \times S^1, \varphi \times \mathrm{id})$ represents $\gamma(y)$. Clearly $\alpha\gamma(y) = y$.

The map β is given as follows: let (W, F) represent an element $z \in \mathscr{S}(N \times I)$ and $f^{\pm} = F|_{\partial^{\pm}W}$. Let M be the result of glueing ∂^+W to ∂^-W via $(f^-)^{-1} \circ f^+$ (Figure 6.17). Then F induces a homotopy equivalence $\hat{F}\colon M \to N \times S^1$ and $\beta(z)$ is defined to be the equivalence class of $(M, \hat{F})$ (Figure 6.18).

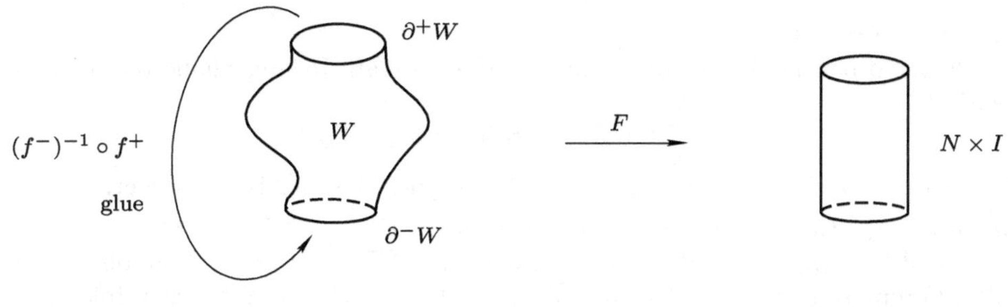

Figure 6.17

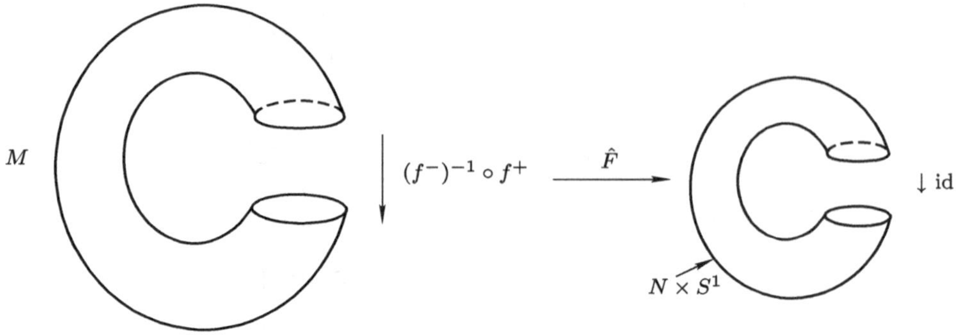

Figure 6.18

Clearly "$\ker \alpha = \operatorname{im} \beta$"; i.e. the preimage of the base point in $\mathscr{S}(N)$ under α equals the image of β.

And another application of the fibering theorem shows that β is one-to-one. □

Now we can calculate $\mathscr{S}(T^n \times I^m)$ by induction on n using this theorem when $n \geq 5$, with $N = T^{n-1} \times I^m$ and noticing that $\mathscr{S}(I^m) = \Theta_m$, the Kervaire-Milnor group of homotopy m-spheres when $m \neq 4$.

The result is (provided $m \geq 5$)

$$\begin{aligned}\mathscr{S}(T^n \times I^m) &= \bigoplus_{i=m}^{n+m} H^i(T^n \times \mathbb{D}^m, \partial; \Theta_i) \\ &= \bigoplus_{j=0}^{n} H^j(T^n; \Theta_{j+m}) \text{ (by Thom isomorphism).}\end{aligned}$$

And if the fibering theorem were true in dimension 4 and 5 (hence in *all* dimensions) we would get that the above formula was true in all dimensions n (but with $m \geq 2$). In particular

$$\mathscr{S}(T^2 \times \mathbb{D}^3) = \bigoplus_{j=0}^{2} H^j(T^2; \Theta_{j+3}) = H^1(T^2; \Theta_4) = \Theta_4 \oplus \Theta_4$$

since $\Theta_3 = \Theta_5 = 0$.

On the other hand, the *correct* answer due to Wall, Hsiang-Shaneson independently is

$$\mathscr{S}(T^2 \times \mathbb{D}^3) = H^0(T^2; \mathbb{Z}_2) = \mathbb{Z}_2.$$

These are inconsistent, since $\mathbb{Z}_2 = \Theta_4 \oplus \Theta_4$ implies that $\sqrt{2}$ is an integer.

Remark 6.57. Here is a proof of the fact that $\mathscr{S}(I^m) = \mathscr{S}(\mathbb{D}^m) = \Theta_m$.

Recall that $\Theta_m = \mathscr{S}(S^m)$. Define a map $\alpha\colon \mathscr{S}(\mathbb{D}^m) \to \mathscr{S}(S^m)$ as follows: let $(\mathscr{D}^m, f)$ represent $x \in \mathscr{S}(\mathbb{D}^m)$, then $f|_{\partial\mathscr{D}^m}\colon \partial\mathscr{D}^m \to \partial\mathbb{D}^m = S^{m-1}$ is a diffeomorphism. Then $\alpha(x) \in \mathscr{S}(S^m)$ is represented by $(\Sigma^m = \mathscr{D}^m \cup_f \mathbb{D}^m, f \cup \mathrm{id})$ (Figure 6.19).

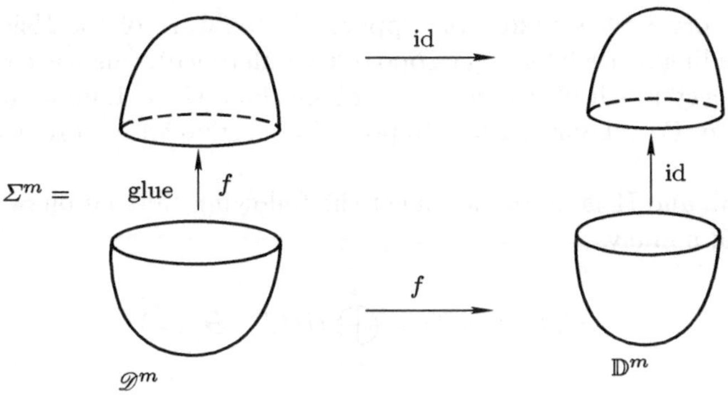

Figure 6.19

The inverse of α is $\beta\colon \mathscr{S}(S^m) \to \mathscr{S}(\mathbb{D}^m)$ defined as follows: let (Σ^m, f) represent $y \in \mathscr{S}(S^m)$, where $f\colon \Sigma^m \to S^m$ is a homotopy equivalence. Let $\mathbb{D}^m \subset \Sigma^m$ be in a small coordinate chart (Figure 6.20).

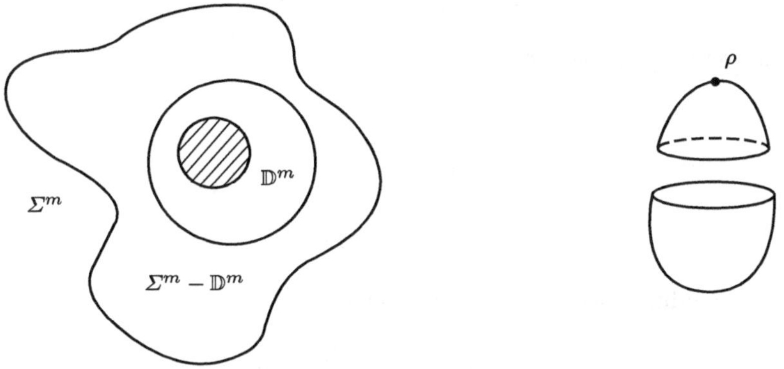

Figure 6.20

We can homotope f so that it is smooth and transverse to the north pole p_N of $S^m = \mathbb{D}^m_+ \cup \mathbb{D}^m_-$. And we can arrange that $f^{-1}(p_N) = \{q\}$ since f is degree 1. Furthermore we can arrange that $\mathbb{D}^m \subset \Sigma^m$ centered at q in a coordinate chart so that $f|_{\mathbb{D}^m} : \mathbb{D}^m \to \mathbb{D}^m$ is a diffeomorphism (i.e. the identity) and $f(\Sigma^m - \mathbb{D}^m)$ is contained in the southern hemisphere. Then $\beta(y)$ is represented by $(\mathscr{D}^m = \Sigma^m - \mathrm{int}(\mathbb{D}^m), f|_{\mathscr{D}^m})$.

What is needed for the Wall, Hsiang-Shaneson calculation is the non-simply connected surgery theory developed by Wall in 1968 to prove this result. Wall had priority over Hsiang-Shaneson, but his published calculation (in his 1971 book [21]) made use of *still unpublished* work of Dennis Sullivan (Characteristic Variety Theorem) while Hsiang-Shaneson did not use this result (they used the fibering theorem in its place). Because the calculation is crucial to the whole Kirby-Siebenmann triangulation theory and Novikov was apparently unaware of the Hsiang-Shaneson paper (Proceeding of 1969 Georgia Topology Conference), this has caused Novikov to publicly reject much of geometric topology done since 1969 as unproven. See Mathematisches Forschungsinstitut Report No.36/2006 for a more detailed discussion.

In fact Wall and Hsiang-Shaneson get the following calculation of $\mathscr{S}(T^n \times \mathbb{D}^m)$ for $n + m \geq 5$; namely,

$$\mathscr{S}(T^n \times \mathbb{D}^m) = \bigoplus_{j=0}^{n} H^j(T^n; \hat{\Theta}_{j+m})$$

where the groups $\hat{\Theta}_j$ differ from Θ_j for only two indices; namely $\hat{\Theta}_3 = \mathbb{Z}_2$ while $\Theta_3 = 0$ and $\hat{\Theta}_4 = 0$ while Θ_4 is still unknown. Otherwise $\hat{\Theta}_i = \Theta_i$.

We next note that the correspondence between $\mathscr{S}(T^n)$ and $\bigoplus_{i \geq 0} H^i(T^n; \hat{\Theta}_i)$ respects finite sheeted covers; i.e. if $p\colon T^n \to T^n$ is a finite sheeted cover (it is an exercise to show that the total space of any finite sheeted cover of T^n is diffeomorphic to T^n), then

$$p^*\colon \mathscr{S}(T^n) \to \mathscr{S}(T^n)$$

defined by $(M^n, f) \mapsto (\hat{M}^n, \hat{f})$, where

$$\begin{array}{ccc} \hat{M}^n & \xrightarrow{\hat{f}} & T^n \\ \downarrow & & \downarrow p \\ M^n & \xrightarrow{f} & T^n \end{array}$$

is the corresponding cover of M^n, corresponds to

$$p^*\colon \bigoplus_{i=0}^{n} H^i(T^n; \hat{\Theta}_i) \to \bigoplus_{i=0}^{n} H^i(T^n; \hat{\Theta}_i).$$

6.8 Siebenmann's example

Now for Siebenmann's example of a closed smooth manifold N^6 homeomorphic but *not* diffeomorphic, hence *not* PL homeomorphic to T^6 by Hirsch-Mazur smoothing theory, we consider the Burghelea map $\beta\colon \mathscr{S}(T^5 \times [0,1]) \to \mathscr{S}(T^6)$. Under the Wall, Hsiang-Shaneson correspondence

$$\mathscr{S}(T^5 \times [0,1]) = H^3(T^5 \times I, \partial; \hat{\Theta}_3) = H^3(T^5 \times S^1, T^5 \times A; \hat{\Theta}_3)$$

and $\mathscr{S}(T^6) = H^3(T^5 \times S^1; \hat{\Theta}_3)$, where A is the left (closed) semi-circle in S^1, β is identified with the map in cohomology induced by the continuous map

$$j\colon (T^5 \times S^1, \emptyset) \to (T^5 \times S^1, T^5 \times A).$$

And by Thom Isomorphism Theorem, $H^3(T^5 \times [0,1], \partial; \hat{\Theta}_3) = H^2(T^5; \hat{\Theta}_3)$.

Let $q_s \colon S^1 \to S^1$ be $q_s(z) = z^s$ and $p_s \colon T^6 \to T^6$ be $p_s = q_s \times \cdots \times q_s \times \mathrm{id}$, where s is a large odd positive integer. Let $x \in \mathscr{S}(T^5 \times [0,1])$ be a fixed non-zero element. And let $y \in \mathscr{S}(T^6)$ be $\beta(x)$. Note that $y \neq 0$ by the Burghelea exact sequence or the exact cohomology sequence. Using the above identification with H^3, we see that $p_s^*(y) = y$.

In more detail, we have a commutative diagram

$$\begin{array}{ccc} T^6 = (T^5 \times S^1, \emptyset) & \xrightarrow{\ j\ } & (T^5 \times S^1, T^5 \times A) \\ \downarrow{\scriptstyle p_s} & & {\scriptstyle p_s}\downarrow \\ (T^5 \times S^1, \emptyset) & \xrightarrow{\ j\ } & (T^5 \times S^1, T^5 \times A) \end{array}$$

and note that $p_s^* \colon H^3(T^6; \mathbb{Z}_2) \to H^3(T^6; \mathbb{Z}_2)$ is the identity by using the Künneth formula and the fact that s is odd. (Note $q_s^* \colon H^1(S^1; \mathbb{Z}_2) \to H^1(S^1; \mathbb{Z}_2)$ is multiplication by s, hence is identity.)

Let $f \colon N^6 \to T^6$ represent y.

Lemma 6.58. *f is homotopic to a homeomorphism.*

Assuming this lemma, then this is Siebenmann's example; i.e. N^6 is *not* diffeomorphic to T^6. Since if it were we could find a diffeomorphism $g \colon N^6 \to T^6$ which is homotopic to f. But this contradicts the fact that $y \neq 0$; i.e. $(N^6, f) \neq (T^6, \mathrm{id})$.

To see that we can find g homotopic to f, one need only observe that any self homotopy equivalence of T^6 is represented by a diffeomorphism. For this recall that two continuous self maps of T^6 are homotopic if and only if they induce the same map on $\pi_1(T^6) = C^6$. But $Aut(C^6) = GL_6(\mathbb{Z})$. Let $B \in GL_6(\mathbb{Z})$, then the map $G \colon T^6 = \mathbb{R}^6/\mathbb{Z}^6 \to \mathbb{R}^6/\mathbb{Z}^6$ defined by $G(v + \mathbb{Z}^6) = Bv + \mathbb{Z}^6$ is a self diffeomorphism inducing B on π_1.

So it remains to prove Lemma 6.58. This depends on the following very important theorem of Cernavskii, one of the first results in controlled topology.

Theorem 6.59 (Cernavskii, Doklady announcement 1968). *The homeomorphism group of a closed manifold M^n is locally contractible. In particular, there exists an $\varepsilon = \varepsilon_M > 0$ such that if $f \colon M^n \to M^n$ is a homeomorphism with $d(f(x), x) \leq \epsilon$ for all $x \in M$, then f is (topologically) isotopic to id_{M^n}.*

Let $F \colon W^6 \to T^5 \times [0,1]$ represent $x \in \mathscr{S}(T^5 \times [0,1])$. Then W^6 is an h-cobordism and hence a cylinder since $Wh(C^5) = 0$. Furthermore $\partial^- W$ is diffeomorphic to T^5. Hence we may assume that $W = T^5 \times [0,1]$ and $F|_{T^5 \times 0} = \mathrm{id}_{T^5}$ while $\varphi = F|_{T^5 \times 1}$ is a self-diffeomorphism of T^5 and F is a homotopy of φ to id_{T^5}. And since $y = \beta(x)$, N^6 is the mapping torus of φ; i.e. $N^6 = (T^5)_\varphi$ and f is induced by F (Figure 6.21).

Furthermore $p_s^*(y)$ is represented by $((T^5)_{\varphi_s}, \hat{F}_s)$ where F_s is the lift of F to the covering space

$$q_s \times \cdots \times q_s \times \mathrm{id} \colon T^5 \times [0,1] \to T^5 \times [0,1]$$

such that $F_s|_{T^5\times 0} = \text{id}$; i.e. the following diagram commutes

$$\begin{array}{ccc} T^5 \times [0,1] & \xrightarrow{F_s} & T^5 \times [0,1] \\ \Big\downarrow{\scriptstyle q_s\times\cdots\times q_s\times \text{id}} & & \Big\downarrow{\scriptstyle q_s\times\cdots\times q_s\times \text{id}} \\ T^5 \times [0,1] & \xrightarrow[F]{} & T^5 \times [0,1] \end{array} \qquad (*)$$

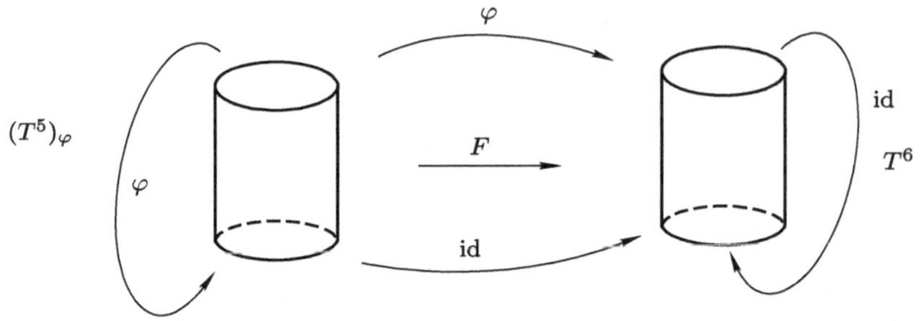

Figure 6.21

Remark 6.60. T^5 is a topological group and $(q_s \times \cdots \times q_s)(z) = z^s$. And $\varphi_s = F_s|_{T^5\times 1}$ while $\hat{F}_s$ is the homotopy equivalence $(T^5)_{\varphi_s} \to (T^5)_{\text{id}} = T^6$ induced by F_s (Figure 6.22).

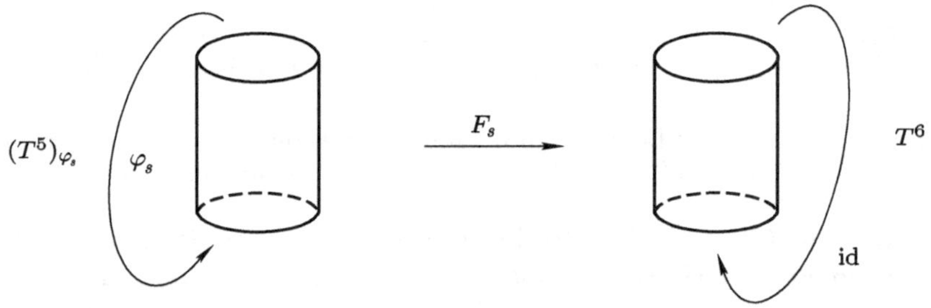

Figure 6.22

Now we come to the crucial fact.

Lemma 6.61. *φ_s is ϵ_{T^5}-close to id_{T^5} if s is large enough.*

Assuming this we complete the proof of Lemma 6.58. Since φ_s is then ϵ_{T^5}-close to id_{T^5}, φ_s is isotopic to id_{T^5} by Cernavskii's Theorem and therefore $(T^5)_{\varphi_s}$ is homeomorphic to $(T^5)_{\text{id}} = T^6$. In fact, let $h_t \colon T^5 \to T^5$ ($t \in [0,1]$) be the isotopy between $h_0 = \text{id}$ and $h_1 = \varphi_s$, then $\hat{h} \colon T^5_{\varphi_s} \to T^5_{\text{id}}$ induced by h_t is a homeomorphism (Figure 6.23).

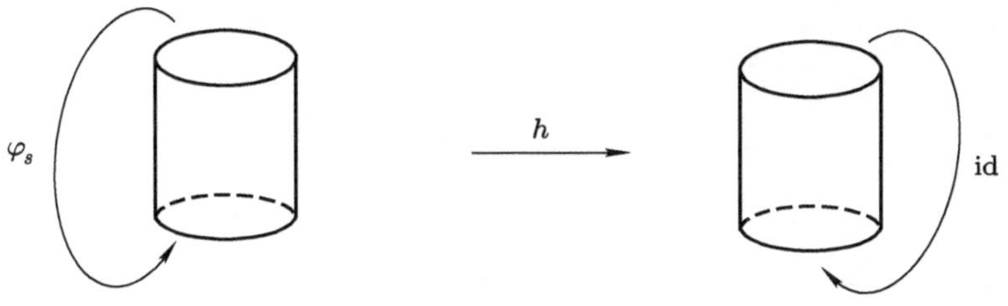

Figure 6.23

Now after composing $\hat{h}$ with a self-homeomorphism (in fact diffeomorphism) of T^6 we get a homeomorphism $\overline{h}\colon (T^5)_{\varphi_s} \to T^6$ which is homotopic to $\hat{F}_s$. But since $y = p_s^*(y)$, there exists a diffeomorphism $\alpha\colon N \to (T^5)_{\varphi_s}$ such that f is homotopic to $\hat{F}_s \circ \alpha$. (Recall that y and $p_s^*(y)$ are represented by (N, f) and $((T^5)_{\varphi_s}, \hat{F}_s)$, respectively.) And since $\hat{F}_s$ is homotopic to $\overline{h}$, we conclude that f is homotopic to $\overline{h} \circ \alpha$, which is a homeomorphism, proving Lemma 6.58 .

So it remains to verify Lemma 6.61.

Note we can assume that F and hence also F_s are both smooth maps. Denote

$$F(x,t) = (\gamma(x,t), \omega(x,t)) \in T^5 \times [0,1]$$

where$(x,t) \in T^5 \times [0,1]$ and likewise $F_s(x,t) = (\hat{\gamma}(x,t), \hat{\omega}(x,t))$. And because of diagram $(*)$ $\hat{\omega}(x,t) = \omega(x^s, t)$ while

$$(\hat{\gamma}(x,t))^s = \gamma(x^s, t). \tag{6.6}$$

We are particularly interested in this equation. Let $E\colon T^5 \to T^5$ be the map $E(z) = z^s$; i.e. $E(z_1, \cdots, z_5) = (z_1^s, \cdots, z_5^s)$. Then $|dE(v)| = s|v|$ for each vector v tangent to T^5. Consider the smooth curves $\hat{\gamma}_x\colon t \mapsto \hat{\gamma}(x,t)$, $\gamma_x\colon t \mapsto \gamma(x,t)$ and $t \mapsto E(\hat{\gamma}(x,t))$. Let $\|\alpha\|$ denote the arc length of a smooth curve α in T^5. Since $E \circ \hat{\gamma}_x = \gamma_{x^s}$ by Eq. (6.6), $\|\hat{\gamma}_x\| = \frac{1}{s}\|\gamma_{x^s}\|$. But the family of numbers $\{\|\gamma_x\| \mid x \in T^5\}$ has a finite upper bound B independent of s, since $\gamma_x(t) = \gamma(x,t)$ has compact domain $T^5 \times [0,1]$. Therefore $\|\hat{\gamma}_x\| \le B/s$ for all $x \in T^5$ and $s \in \mathbb{Z}^+$. Hence $d(\hat{\gamma}_x(1), \hat{\gamma}_x(0)) \le B/s$. But $\hat{\gamma}_x(0) = \hat{\gamma}(x,0) = x$ while $\hat{\gamma}(x,1) = \varphi_s(x)$. Therefore $d(\varphi_s(x), x) \le B/s$. This proves Lemma 6.61 when $B/s < \varepsilon$.

References

[1] Hyman Bass, Algebraic K-theory, W. A. Benjamin, Inc., New York-Amsterdam 1968.

[2] Raoul Bott and John W. Milnor, *On the parallelizability of the spheres*, Bull. Amer. Math. Soc. 64 (1958), 87–89.

[3] William Browder, Surgery on Simply-connected Manifolds, Ergebnisse der Mathematik und ihrer Grenzgebiete, Band 65, Springer-Verlag, New York-Heidelberg, 1972.

[4] Thomas Farrell and J. B. Wagoner, *Infinite matrices in algebraic K-theory and topology*, Comm. Math. Helv. 47 (1972), no. 1, 474–501.

[5] Michael Freedman, *The topology of four-dimensional manifolds*, J. Differential Geom. 17 (1982), no. 3, 357–453.

[6] Martin. Hertweck, *A counterexample to the isomorphism problem for integral group rings*, Ann. of Math. (2) 154 (2001), no. 1, 115–138.

[7] Morris W. Hirsch, Differential Topology, Graduate Texts in Mathematics 33, Springer-Verlag, New York-Heidelberg, 1976.

[8] Morris W. Hirsch and Barry Mazur, Smoothings of Piecewise Linear Manifolds, Annals of Mathematics Studies 80, Princeton University Press, Princeton, N. J.; University of Tokyo Press, Tokyo, 1974.

[9] Michel A. Kervaire and John W. Milnor, *Bernoulli numbers, homotopy groups, and a theorem of Rohlin*, Proc. Internat. Congress Math., 1958, Cambridge Univ. Press, New York, (1960), 454–458.

[10] Michel A. Kervaire and John W. Milnor, *Groups of homotopy spheres.* I, Ann. of Math. (2) 77 (1963), 504–537.

[11] Robion C. Kirby and Laurence C. Siebenmann, Foundational Essays on Topological Manifolds, Smoothings, and Triangulations, Annals of Mathematics Studies 88, Princeton University Press, Princeton, N.J.; University of Tokyo Press, Tokyo, 1977.

[12] Barry Mazur, *Stable equivalence of differentiable manifolds*, Bull. Amer. Math. Soc. 67 (1961), 377–384.

[13] John W. Milnor, Lectures on the h-Cobordism Theorem, Princeton University Press, Princeton, N.J., 1965.

[14] John W. Milnor, Morse Theory, Annals of Mathematics Studies 51, Princeton University Press, Princeton, N.J., 1963.

[15] John W. Milnor, *Whitehead torsion*, Bull. AMS 72 (1966), no. 3, 356–426.

[16] John W. Milnor, *Two complexes which are homeomorphic but combinatorially distinct*, Ann. of Math. (2) 74 (1961), 575–590.

[17] John W. Milnor and James D. Stasheff, Characteristic Classes, Annals of Mathematics Studies 76, Princeton University Press, Princeton, N. J.; University of Tokyo Press, Tokyo, 1974.

[18] Jonathan Rosenberg, Algebraic K-Theory and Its Applications, Graduate Texts in Mathematics 147, Springer-Verlag, New York, 1994.

[19] Norman Steenrod, Homology of Regular Cell Complexes, Princeton University Press, Preliminary Lectures Notes Series (noted by Cooke and Finney), 1968.

[20] Norman Steenrod, The Topology of Fibre Bundles, Princeton Mathematical Series 14, Princeton University Press, Princeton, N. J., 1951.

[21] C. T. C. Wall, Surgery on Compact Manifolds, London Mathematical Society Monographs 1, Academic Press, London-New York, 1970.

Index

Made in United States
Troutdale, OR
01/28/2026

46245630R10077